"Casts a golden glow on family life...Re
exciting as reading a thriller—though the on
child's entry into consciousness....Her b
alive."

"This sunny memoir...reminds one of an Impressionist painting; her memories shimmer on the page." —Cyra McFadden, *Los Angeles Times*

"[Annie Dillard] is one of those people who seem to be more fully alive than most of us, more nearly wide-awake than human beings generally get to be." —Noel Perrin, *New York Times*

"As I read *An American Childhood*, I [had] the astonishing experience of seeing my neighborhood, my very childhood, unroll before my eyes in the pages of someone else's book." —*USA Today*

"Annie Dillard has been twice-blessed—by literate and prosperous parents, who gave her unconditional love and freedom of mind and person, and by her own extraordinary gifts of observation and language. The reader of *An American Childhood* reaps the fruit of these blessings." —Hilma Wolitzer, *Newsday*

"A catalogue of loves lovingly told...This delighted exploration of the world of books is by far the most enjoyable thing in *An American Childhood* and, in its modest way, a classic love story." —*Washington Post*

"Marvelous." —*Kansas City Star*

"*An American Childhood* might be described as a metaphysical memoir—closer in spirit to Rousseau or St. Augustine than to most contemporary memoirists....Her writing celebrates and revels in its surprising loops and leaps of insight the value of the scrupulously examined life." —*Providence Journal*

"Dillard is a natural stylist with a flair and keen love for words. We've seen it in her *Pilgrim at Tinker Creek*, *Living by Fiction*, and now this, her most luminous work....*An American Childhood's* penultimate chapter is literature on high beam. Her memories rush and bubble up, a kind of waterfall of language and feeling, in summation....Annie Dillard's *An American Childhood* is a glorious, exultant book. You must read it." —*Columbus Dispatch*

"A rare treat, an autobiography you'll want to read more than once...When she...reawaken[s] your own joy, you can't help but be grateful." —*San Jose Mercury-News*

AN AMERICAN CHILDHOOD

BY THE SAME AUTHOR

Encounters with Chinese Writers

Teaching a Stone to Talk

Living by Fiction

Holy the Firm

Pilgrim at Tinker Creek

Tickets for a Prayer Wheel

AN AMERICAN CHILDHOOD

ANNIE DILLARD

Parts of this book have appeared, in different form, in the *New York Times Magazine*, *American Heritage Magazine*, and the *New York Times Book Review*.

A hardcover edition of this book was published in 1987 by Harper & Row, Publishers.

First PERENNIAL LIBRARY edition published 1988.

Designed by Ruth Bornschlegel

Library of Congress Cataloging-in-Publication Data

Dillard, Annie.
An American childhood.

"Perennial Library."

1. Dillard, Annie—Biography—Youth. 2. Authors, American—20th century—Biography. 3. Pittsburgh (Pa.)—Social life and customs. I. Title.
PS3554.I398Z464 1988 818'.5409 [B] 87-45042
ISBN 0-06-091518-8 (pbk.)

---

03 02 01 00 RRD(H) 40 39 38 37 36 35 34 33 32 31

for my parents
PAM LAMBERT DOAK
and
FRANK DOAK

A grant from the John
Simon Guggenheim Memorial Foundation
aided this work.

I have loved, O Lord, the beauty
of thy house and the place
where dwelleth thy glory.

PSALM 26

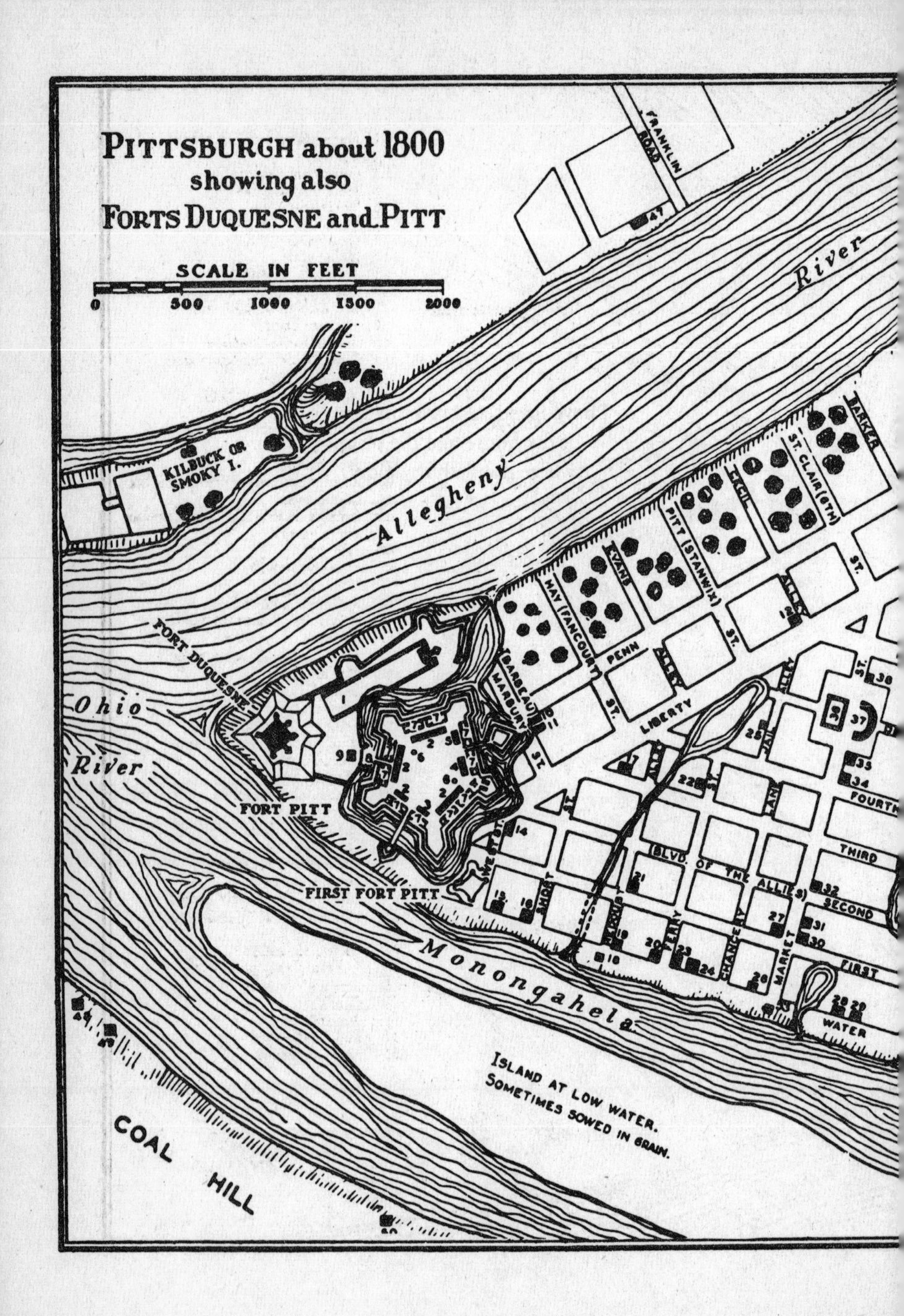
PITTSBURGH about 1800
showing also
FORTS DUQUESNE and PITT
SCALE IN FEET
0
500
1000
1500
2000
FRANKLIN ROAD
River
KILBUCK OR SMOKY I.
Allegheny
FORT DUQUESNE
Ohio
River
FORT PITT
FIRST FORT PITT
Monongahela
ISLAND AT LOW WATER.
SOMETIMES SOWED IN GRAIN.
COAL
HILL
MARBURY
MAY (FANCOURT)
EVANS
PITT (STANWIX)
ST. CLAIR (6TH)
BARKER
PENN
ST.
LIBERTY
FOURTH
THIRD
(BLVD. OF THE ALLIES)
SECOND
FIRST
WATER
SHORT
FERRY
CHANCERY
MARKET

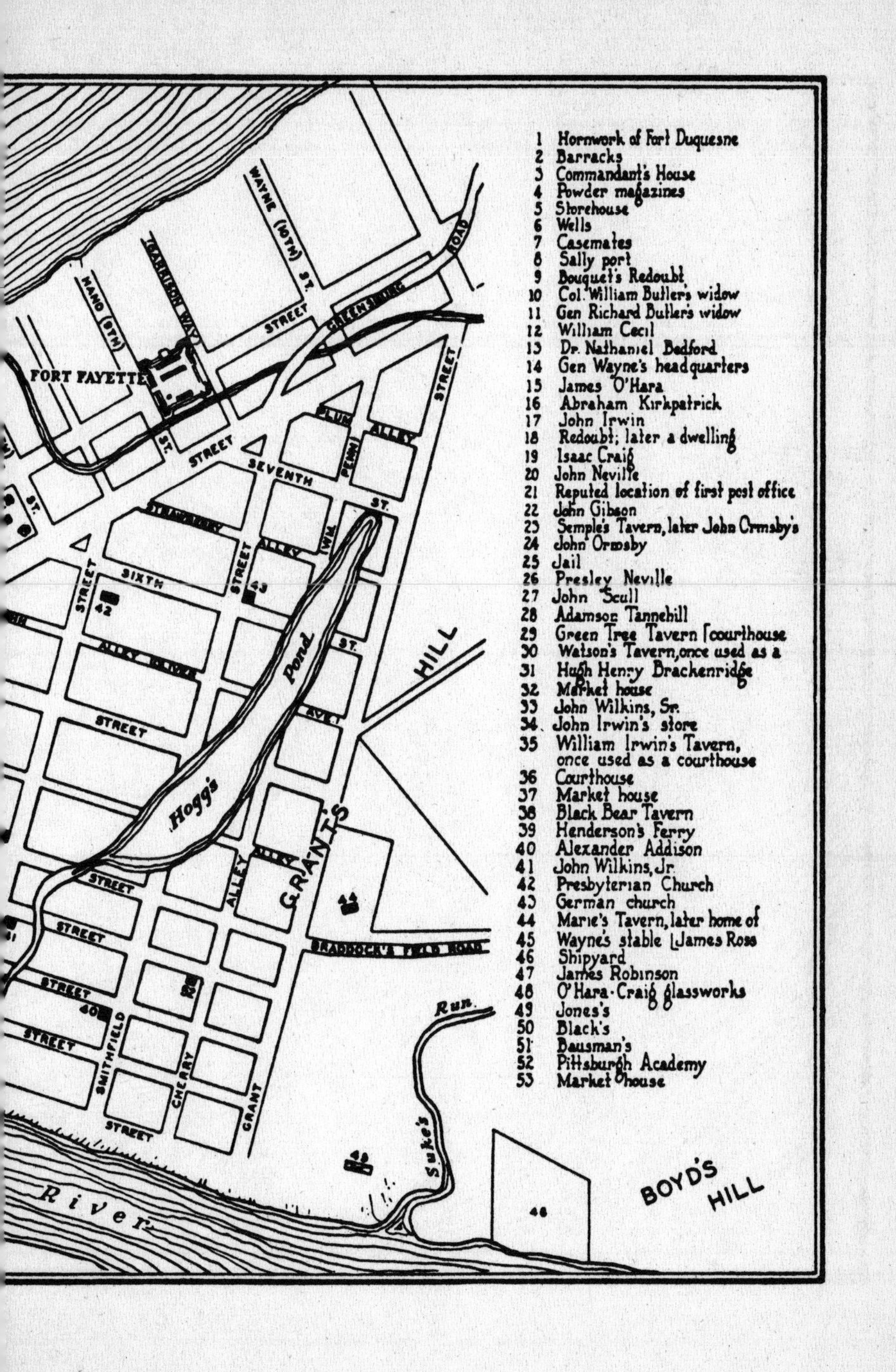
1 Hornwork of Fort Duquesne
2 Barracks
3 Commandant's House
4 Powder magazines
5 Storehouse
6 Wells
7 Casemates
8 Sally port
9 Bouquet's Redoubt
10 Col. William Butler's widow
11 Gen Richard Butler's widow
12 William Cecil
13 Dr. Nathaniel Bedford
14 Gen Wayne's headquarters
15 James O'Hara
16 Abraham Kirkpatrick
17 John Irwin
18 Redoubt; later a dwelling
19 Isaac Craig
20 John Neville
21 Reputed location of first post office
22 John Gibson
23 Semple's Tavern, later John Ormsby's
24 John Ormsby
25 Jail
26 Presley Neville
27 John Scull
28 Adamson Tannehill
29 Green Tree Tavern
30 Watson's Tavern, once used as a courthouse
31 Hugh Henry Brackenridge
32 Market house
33 John Wilkins, Sr.
34 John Irwin's store
35 William Irwin's Tavern, once used as a courthouse
36 Courthouse
37 Market house
38 Black Bear Tavern
39 Henderson's Ferry
40 Alexander Addison
41 John Wilkins, Jr.
42 Presbyterian Church
43 German church
44 Marie's Tavern, later home of James Ross
45 Wayne's stable
46 Shipyard
47 James Robinson
48 O'Hara-Craig glassworks
49 Jones's
50 Black's
51 Bausman's
52 Pittsburgh Academy
53 Market house
WAYNE (10TH) ST.
HAND (9TH)
STREET
GREENSBURG
ROAD
FORT FAYETTE
ST.
STREET
SEVENTH
PLUM
ALLEY
PENN
STREET
ST.
STRAWBERRY
ALLEY
WM.
SIXTH
STREET
HILL
Pond
Hogg's
AVE.
GRANT'S
BRADDOCK'S FIELD ROAD
SMITHFIELD
CHERRY
GRANT
Run
Suke's
River
BOYD'S
HILL

Prologue

WHEN EVERYTHING ELSE HAS GONE from my brain—the President's name, the state capitals, the neighborhoods where I lived, and then my own name and what it was on earth I sought, and then at length the faces of my friends, and finally the faces of my family—when all this has dissolved, what will be left, I believe, is topology: the dreaming memory of land as it lay this way and that.

I will see the city poured rolling down the mountain valleys like slag, and see the city lights sprinkled and curved around the hills' curves, rows of bonfires winding. At sunset a red light like housefires shines from the narrow hillside windows; the houses' bricks burn like glowing coals.

The three wide rivers divide and cool the mountains. Calm old bridges span the banks and link the hills. The Allegheny River flows in brawling from the north, from near the shore of Lake Erie, and from Lake Chautauqua in New York and eastward. The Monongahela River flows in shallow and slow from the south, from West Virginia. The Allegheny and the Monongahela meet and form the westward-wending Ohio.

Where the two rivers join lies an acute point of flat land from which rises the city. The tall buildings rise lighted to their tips. Their lights illumine other buildings' clean sides, and illumine the narrow city canyons below, where people move, and shine reflected red and white at night from the black waters.

When the shining city, too, fades, I will see only those forested mountains and hills, and the way the rivers lie flat and moving among them, and the way the low land lies wooded among them, and the blunt mountains rise in darkness from the rivers' banks, steep from the rugged south and rolling from the north, and from farther, from the inclined eastward plateau where the high ridges begin to run so long north and south unbroken that to get around them you practically have to navigate Cape Horn.

In those first days, people said, a squirrel could run the long length of Pennsylvania without ever touching the ground. In those first days, the woods were white oak and chestnut, hickory, maple, sycamore, walnut, wild ash, wild plum, and white pine. The pine grew on the ridgetops where the mountains' lumpy spines stuck up and their skin was thinnest.

The wilderness was uncanny, unknown. Benjamin Franklin had already invented his stove in Philadelphia by 1753, and Thomas Jefferson was a schoolboy in Virginia; French soldiers had been living in forts along Lake Erie for two generations. But west of the Alleghenies in western Pennsylvania, there was not even a settlement, not even a cabin. No Indians lived there, or even near there.

Wild grapevines tangled the treetops and shut out the sun. Few songbirds lived in the deep woods. Bright Carolina parakeets—red, green, and yellow—nested in the dark forest. There were ravens then, too. Woodpeckers rattled the big trees' trunks, ruffed grouse whirred their tail feathers in the fall, and every long once in a while a nervous gang of empty-headed turkeys came hustling and kicking through the leaves—but no one heard any of this, no one at all.

In 1753, young George Washington surveyed for the English this point of land where rivers met. To see the forest-blurred lay of the land, he rode his horse to a ridgetop and climbed a tree. He judged it would make a good spot for a fort. And an English fort it became, and a depot for Indian traders to the Ohio country, and later a French fort and way station to New Orleans.

But it would be another ten years before any settlers lived

there on that land where the rivers met, lived to draw in the flowery scent of June rhododendrons with every breath. It would be another ten years before, for the first time on earth, tall men and women lay exhausted in their cabins, sleeping in the sweetness, worn out from planting corn.

In 1955, when I was ten, my father's reading went to his head.

My father's reading during that time, and for many years before and after, consisted for the most part of *Life on the Mississippi*. He was a young executive in the old family firm, American Standard; sometimes he traveled alone on business. Traveling, he checked into a hotel, found a bookstore, and chose for the night's reading, after what I fancy to have been long deliberation, yet another copy of *Life on the Mississippi*. He brought all these books home. There were dozens of copies of *Life on the Mississippi* on the living-room shelves. From time to time, I read one.

Down the Mississippi hazarded the cub riverboat pilot, down the Mississippi from St. Louis to New Orleans. His chief, the pilot Mr. Bixby, taught him how to lay the boat in her marks and dart between points; he learned to pick a way fastidiously inside a certain snag and outside a shifting shoal in the black dark; he learned to clamber down a memorized channel in his head. On tricky crossings the leadsmen sang out the soundings, so familiar I seemed to have heard them the length of my life: "Mark four! . . . Quarter-less-four! . . . Half three! . . . Mark three! . . . Quarter-less . . ." It was an old story.

When all this reading went to my father's head, he took action. From Pittsburgh he went down the river. Although no one else that our family knew kept a boat on the Allegheny River, our father did, and now he was going all the way with it. He quit the firm his great-grandfather had founded a hundred years earlier down the river at his family's seat in Louisville, Kentucky; he sold his own holdings in the firm. He was taking off for New Orleans.

New Orleans was the source of the music he loved: Dixieland jazz, O Dixieland. In New Orleans men would blow it in the air and beat it underfoot, the music that hustled and snapped, the music whose zip matched his when he was a man-about-town at home in Pittsburgh, working for the family firm; the music he tapped his foot to when he was a man-about-town in New York for a few years after college working for the family firm by day and by night hanging out at Jimmy Ryan's on Fifty-second Street with Zutty Singleton, the black drummer who befriended him, and the rest of the house band. A certain kind of Dixieland suited him best. They played it at Jimmy Ryan's, and Pee Wee Russell and Eddie Condon played it too—New Orleans Dixieland chilled a bit by its journey up the river, and smoothed by its sojourns in Chicago and New York.

Back in New Orleans where he was headed they would play the old stuff, the hot, rough stuff—bastardized for tourists maybe, but still the big and muddy source of it all. Back in New Orleans where he was headed the music would smell like the river itself, maybe, like a thicker, older version of the Allegheny River at Pittsburgh, where he heard the music beat in the roar of his boat's inboard motor; like a thicker, older version of the wide Ohio River at Louisville, Kentucky, where at his family's summer house he'd spent his boyhood summers mucking about in boats.

Getting ready for the trip one Saturday, he roamed around our big brick house snapping his fingers. He had put a record on: Sharkey Bonano, "Li'l Liza Jane." I was reading Robert Louis Stevenson on the sunporch: *Kidnapped*. I looked up from my book and saw him outside; he had wandered out to the lawn and was standing in the wind between the buckeye trees and looking up at what must have been a small patch of wild sky. Old Low-Pockets. He was six feet four, all lanky and leggy; he had thick brown hair and shaggy brows, and a mild and dreamy expression in his blue eyes.

When our mother met Frank Doak, he was twenty-seven: witty, boyish, bookish, unsnobbish, a good dancer. He had grown up an only child in Pittsburgh, attended Shady Side Academy, and Washington and Jefferson College in Pennsylvania, where he studied history. He was a lapsed Presbyterian and a believing Republican. "Books make the man," read the blue bookplate in all his books. "Frank Doak." The bookplate's woodcut showed a square-rigged ship under way in a steep following sea. Father had hung around jazz in New York, and halfheartedly played the drums; he had smoked marijuana, written poems, begun a novel, painted in oils, imagined a career as a riverboat pilot, and acted for more than ten seasons in amateur and small-time professional theater. At American Standard, Amstan Division, he was the personnel manager.

But not for long, and never again; Mother told us he was quitting to go down the river. I was sorry he'd be leaving the Manufacturers' Building downtown. From his office on the fourteenth floor, he often saw suicides, which he reported at dinner. The suicides grieved him, but they thrilled us kids. My sister Amy was seven.

People jumped from the Sixth Street bridge into the Allegheny River. Because the bridge was low, they shinnied all the way up the steel suspension cables to the bridge towers before they jumped. Father saw them from his desk in silhouette, far away. A man vigorously climbed a slanting cable. He slowed near the top, where the cables hung almost vertically; he paused on the stone tower, seeming to sway against the sky, high over the bridge and the river below. Priests, firemen, and others—presumably family members or passersby—gathered on the bridge. In about half the cases, Father said, these people talked the suicide down. The ones who jumped kicked off from the tower so they'd miss the bridge, and fell tumbling a long way down.

Pittsburgh was a cheerful town, and had far fewer suicides than most other cities its size. Yet people jumped so often that Father and his colleagues on the fourteenth floor had a betting pool going. They guessed the date and time of day the next jumper would appear. If a man got talked down before he jumped, he still counted for the betting pool, thank God; no manager of American Standard ever wanted to hope, even in the smallest part of himself, that the fellow would go ahead and jump. Father said he and the other men used to gather at the biggest window and holler, "No! Don't do it, buddy, don't!" Now he was leaving American Standard to go down the river, and he was a couple of bucks in the hole.

While I was reading *Kidnapped* on this Saturday morning, I heard him come inside and roam from the kitchen to the pantry to the bar, to the dining room, the living room, and the sunporch, snapping his fingers. He was snapping the fingers of both hands, and shaking his head, to the record—"Li'l Liza Jane"—the sound that was beating, big and jivey, all over the house. He walked lightly, long-legged, like a soft-shoe hoofer barely in touch with the floor. When he played the drums, he played lightly, coming down soft with the steel brushes that sounded like a Slinky falling, not making the beat but just sizzling along with it. He wandered into the sunporch, unseeing; he was snapping his fingers lightly, too, as if he were feeling between them a fine layer of Mississippi silt. The big buckeyes outside the glass sunporch walls were waving.

A week later, he bade a cheerful farewell to us—to Mother, who had encouraged him, to us oblivious daughters, ten and seven, and to the new baby girl, six months old. He loaded his twenty-four-foot cabin cruiser with canned food, pushed off from the dock of the wretched boat club that Mother hated, and pointed his bow downstream, down the Allegheny River. From there it was only a few miles to the Ohio River at Pittsburgh's point, where the Monongahela came in. He wore on westward down the Ohio; he watched West Virginia float past his port bow and Ohio past his starboard. It was 138 river miles to New Martinsville, West Virginia, where he lingered for some races. Back on the move, he tied up nights at club docks he'd seen on the charts; he poured himself water for drinks from dockside hoses. By day he rode through locks, twenty of them in all. He conversed with the lockmasters, those lone men who paced silhouetted in overalls on the concrete lock-chamber walls and threw the big switches that flooded or drained the locks: "Hello, up there!" "So long, down there!"

He continued down the river along the Kentucky border with Ohio, bumping down the locks. He passed through Cincinnati. He moved along down the Kentucky border with Indiana. After 640 miles of river travel, he reached Louisville, Kentucky. There he visited relatives at their summer house on the river.

It was a long way to New Orleans, at this rate another couple of months. He was finding the river lonesome. It got dark too early. It was September; people had abandoned their pleasure boats for the season; their children were back in school. There were no old salts on the docks talking river talk. People weren't so friendly as they were in Pittsburgh. There was no music except the dreary yacht-club jukeboxes playing "How Much Is That Doggie in the Window?" Jazz had come up the river once and for all; it wasn't still coming, he couldn't hear it across the water at night rambling and blowing and banging along high and tuneful, sneaking upstream to Chicago to get educated. He wasn't free so much as loose. He was living alone on beans in a boat and having witless conversations with lockmasters. He mailed out sad postcards.

From phone booths all down the Ohio River he talked to Mother. She told him that she was lonesome, too, and that three children—maid and nanny or no—were a handful. She said, further, that people were starting to talk. She knew Father couldn't bear people's talking. For all his dreaminess, he prized respectability above all; it was our young mother, whose circumstances bespoke such dignity, who loved to shock the world. After only six weeks, then—on the Ohio River at Louisville—he sold the boat and flew home.

I was just waking up then, just barely. Other things were changing. The highly entertaining new baby, Molly, had taken up residence in a former guest room. The great outer world hove into view and began to fill with things that had apparently been there all along: mineralogy, detective work, lepidopterology, ponds and streams, flying, society. My younger sister Amy and I were to start at private school that year: the Ellis School, on Fifth Avenue. I would start dancing school.

Children ten years old wake up and find themselves here, discover themselves to have been here all along; is this sad? They wake like sleepwalkers, in full stride; they wake like people brought back from cardiac arrest or from drowning: *in medias res,* surrounded by familiar people and objects, equipped with a hundred skills. They know the neighborhood, they can read and write English, they are old hands at the commonplace mysteries, and yet they feel themselves to have just stepped off the boat, just converged with their bodies, just flown down from a trance, to lodge in an eerily familiar life already well under way.

I woke in bits, like all children, piecemeal over the years. I discovered myself and the world, and forgot them, and discovered them again. I woke at intervals until, by that September when Father went down the river, the intervals of waking tipped the scales, and I was more often awake than not. I noticed this process of waking, and predicted with terrifying logic that one of these years not far away I would be awake continuously and never slip back, and never be free of myself again.

Consciousness converges with the child as a landing tern touches the outspread feet of its shadow on the sand: precisely, toe hits toe. The tern folds its wings to sit; its shadow dips and spreads over the sand to meet and cup its breast.

Like any child, I slid into myself perfectly fitted, as a diver meets her reflection in a pool. Her fingertips enter the fingertips on the water, her wrists slide up her arms. The diver wraps herself in her reflection wholly, sealing it at the toes, and wears it as she climbs rising from the pool, and ever after.

I never woke, at first, without recalling, chilled, all those other waking times, those similar stark views from similarly lighted precipices: dizzying precipices from which the distant, glittering world revealed itself as a brooding and separated scene—and so let slip a queer implication, that I myself was both observer and observable, and so a possible object of my own humming awareness. Whenever I stepped into the porcelain bathtub, the bath's hot water sent a shock traveling up my bones. The skin on my arms pricked up, and the hair rose on the back of my skull. I saw my own firm foot press the tub, and the pale shadows waver over it, as if I were looking down from the sky and remembering this scene forever. The skin on my face tightened, as it had always done whenever I stepped into the tub, and remembering it all drew a swinging line, loops connecting the dots, all the way back. You again.

Part One

THE STORY STARTS BACK IN 1950, when I was five.

Oh, the great humming silence of the empty neighborhoods in those days, the neighborhoods abandoned everywhere across continental America—the city residential areas, the new "suburbs," the towns and villages on the peopled highways, the cities, towns, and villages on the rivers, the shores, in the Rocky and Appalachian mountains, the piedmont, the dells, the bayous, the hills, the Great Basin, the Great Valley, the Great Plains—oh, the silence!

For every morning the neighborhoods emptied, and all vital activity, it seemed, set forth for parts unknown.

The men left in a rush: they flung on coats, they slid kisses at everybody's cheeks, they slammed house doors, they slammed car doors; they ground their cars' starters till the motors caught with a jump.

And the Catholic schoolchildren left in a rush; I saw them from our dining-room windows. They burst into the street buttoning their jackets; they threw dry catalpa pods at the stop sign and at each other. They hugged their brown-and-tan workbooks to them, clumped and parted, and proceeded toward St. Bede's church school almost by accident.

The men in their oval, empty cars drove slowly among the schoolchildren. The boys banged the cars' fenders with their hands, with their jackets' elbows, or their books. The men in cars inched among the children; they edged around corners and vanished from sight. The waving knots of children zigzagged and hollered up the street and vanished from sight. And inside all the forgotten houses in all the abandoned neighborhoods, the day of silence and waiting had begun.

The war was over. People wanted to settle down, apparently, and calmly blow their way out of years of rationing. They wanted to bake sugary cakes, burn gas, go to church together, get rich, and make babies.

I had been born at the end of April 1945, on the day Hitler died; Roosevelt had died eighteen days before. My father had been 4-F in the war, because of a collapsing lung—despite his repeated and chagrined efforts to enlist. Now—five years after V-J Day—he still went out one night a week as a volunteer to the Civil Air Patrol; he searched the Pittsburgh skies for new enemy bombers. By day he worked downtown for American Standard.

Every woman stayed alone in her house in those days, like a coin in a safe. Amy and I lived alone with our mother most of the day. Amy was three years younger than I. Mother and Amy and I went our separate ways in peace.

The men had driven away and the schoolchildren had paraded out of sight. Now a self-conscious and stricken silence overtook the neighborhood, overtook our white corner house and myself inside. "Am I living?" In the kitchen I watched the unselfconscious trees through the screen door, until the trees' autumn branches like fins waved away the silence. I forgot myself, and sank into dim and watery oblivion.

A car passed. Its rush and whine jolted me from my blankness. The sound faded again and I faded again down into my hushed brain until the icebox motor kicked on and prodded me awake. "You are living," the icebox motor said. "It is morning, morning, here in the kitchen, and you are in it," the icebox motor said, or the dripping faucet said, or any of the hundred other noisy things that only children can't stop hearing. Cars started, leaves rubbed, trucks' brakes whistled, sparrows peeped. Whenever it rained, the rain spattered, dripped, and ran, for the entire length of the shower, for the entire length of days-long rains, until we children were almost insane from hearing it rain because we couldn't stop hearing it rain. "Rinso white!" cried the man on the radio. "Rinso blue." The silence, like all silences, was made poignant and distinct by its sounds.

What a marvel it was that the day so often introduced itself with a firm footfall nearby. What a marvel it was that so many times a day the world, like a church bell, reminded me to recall and contemplate the durable fact that I was here, and had awakened once more to find myself set down in a going world.

In the living room the mail slot clicked open and envelopes clattered down. In the back room, where our maid, Margaret Butler, was ironing, the steam iron thumped the muffled ironing board and hissed. The walls squeaked, the pipes knocked, the screen door trembled, the furnace banged, and the radiators clanged. This was the fall the loud trucks went by. I sat mindless and eternal on the kitchen floor, stony of head and solemn, playing with my fingers. Time streamed in full flood beside me on the kitchen floor; time roared raging beside me down its swollen banks; and when I woke I was so startled I fell in.

Who could ever tire of this heart-stopping transition, of this breakthrough shift between seeing and knowing you see, between being and knowing you be? It drives you to a life of concentration, it does, a life in which effort draws you down so very deep that when you surface you twist up exhilarated with a yelp and a gasp.

Who could ever tire of this radiant transition, this surfacing to awareness and this deliberate plunging to oblivion—the theater curtain rising and falling? Who could tire of it when the sum of those moments at the edge—the conscious life we so dread losing—is all we have, the gift at the moment of opening it?

Six xylophone notes chimed evenly from the radio in the back room where Margaret was ironing, and then seven xylophone notes chimed. With carefully controlled emotion, a radio woman sang:

What will the weather be?
Tell us, Mister Weather Man.

Mother picked up Amy, who was afraid of the trucks. She called the painters on the phone; it was time to paint the outside trim again. She ordered groceries on the phone. Larry, from Lloyd's Market, delivered. He joked with us in the kitchen while Mother unpacked the groceries' cardboard box.

I wandered outside. It was afternoon. No cars passed on the empty streets; no people passed on the empty sidewalks. The brick houses, the frame and stucco houses, white and red behind their high hedges, were still. A small woman appeared at the far, high end of the street, in silhouette against the sky; she pushed a black baby carriage tall and chromed as a hearse. The leaves in the Lombardy poplars were turning brown.

"Lie on your back," my mother said. She was kind, imaginative. She had joined me in one of the side yards. "Look at the clouds and figure out what they look like. A hat? I see a camel."

Must I? Could this be anybody's idea of something worth doing?

I was hoping the war would break out again, here. I was hoping the streets would fill and I could shoot my cap gun at people instead of at mere sparrows. My project was to ride my swing all around, over the top. I bounced a ball against the house; I fired gravel bits from an illegal slingshot Mother gave me. Sometimes I looked at the back of my hand and tried to memorize it. Sometimes I dreamed of a coal furnace, a blue lake, a redheaded woodpecker who turned into a screeching hag. Sometimes I sang uselessly in the yard, "Blithar, blithar, blithar, blithar."

It rained and it cleared and I sent Popsicle sticks and twigs down the gritty rivulet below the curb. Soon the separated neighborhood trees lost their leaves, one by one. On Saturday afternoons I watched the men rake leaves into low heaps at the curb. They tried to ignite the heaps with matches. At length my father went into the house and returned with a yellow can of lighter fluid. The daylight ended early, before all the men had burned all their leaves.

It snowed and it cleared and I kicked and pounded the snow. I roamed the darkening snowy neighborhood, oblivious. I bit and crumbled on my tongue the sweet, metallic worms of ice that had formed in rows on my mittens. I took a mitten off to fetch some wool strands from my mouth. Deeper the blue shadows grew on the sidewalk snow, and longer; the blue shadows joined and spread upward from the streets like rising water. I walked wordless and unseeing, dumb and sunk in my skull, until—what was that?

The streetlights had come on—yellow, bing—and the new light woke me like noise. I surfaced once again and saw: it was winter now, winter again. The air had grown blue dark; the skies were shrinking; the streetlights had come on; and I was here outside in the dimming day's snow, alive.

THE INTERIOR LIFE is often stupid. Its egoism blinds it and deafens it; its imagination spins out ignorant tales, fascinated. It fancies that the western wind blows on the Self, and leaves fall at the feet of the Self for a reason, and people are watching. A mind risks real ignorance for the sometimes paltry prize of an imagination enriched. The trick of reason is to get the imagination to seize the actual world—if only from time to time.

When I was five, growing up in Pittsburgh in 1950, I would not go to bed willingly because something came into my room. This was a private matter between me and it. If I spoke of it, it would kill me.

Who could breathe as this thing searched for me over the very corners of the room? Who could ever breathe freely again? I lay in the dark.

My sister Amy, two years old, was asleep in the other bed. What did she know? She was innocent of evil. Even at two she composed herself attractively for sleep. She folded the top sheet tidily under her prettily outstretched arm; she laid her perfect head lightly on an unwrinkled pillow, where her thick curls spread evenly in rays like petals. All night long she slept smoothly in a series of pleasant and serene, if artificial-looking, positions, a faint smile on her closed lips, as if she were posing for an ad for sheets. There was no messiness in her, no roughness for things to cling to, only a charming and charmed innocence that seemed then to protect her, an innocence I needed but couldn't muster. Since Amy was asleep, furthermore, and since when I needed someone most I was afraid to stir enough to wake her, she was useless.

I lay alone and was almost asleep when the damned thing entered the room by flattening itself against the open door and sliding in. It was a transparent, luminous oblong. I could see the door whiten at its touch; I could see the blue wall turn pale where it raced over it, and see the maple headboard of Amy's bed glow. It was a swift spirit; it was an awareness. It made noise. It had two joined parts, a head and a tail, like a Chinese dragon. It found the door, wall, and headboard; and it swiped them, charging them with its luminous glance. After its fleet, searching passage, things looked the same, but weren't.

I dared not blink or breathe; I tried to hush my whooping blood. If it found another awareness, it would destroy it.

Every night before it got to me it gave up. It hit my wall's corner and couldn't get past. It shrank completely into itself and vanished like a cobra down a hole. I heard the rising roar it made when it died or left. I still couldn't breathe. I knew—it was the worst fact I knew, a very hard fact—that it could return again alive that same night.

Sometimes it came back, sometimes it didn't. Most often, restless, it came back. The light stripe slipped in the door, ran searching over Amy's wall, stopped, stretched lunatic at the first corner, raced wailing toward my wall, and vanished into the second corner with a cry. So I wouldn't go to bed.

It was a passing car whose windshield reflected the corner streetlight outside. I figured it out one night.

Figuring it out was as memorable as the oblong itself. Figuring it out was a long and forced ascent to the very rim of being, to the membrane of skin that both separates and connects the inner life and the outer world. I climbed deliberately from the depths like a diver who releases the monster in his arms and hauls himself hand over hand up an anchor chain till he meets the ocean's sparkling membrane and bursts through it; he sights the sunlit, becalmed hull of his boat, which had bulked so ominously from below.

I recognized the noise it made when it left. That is, the

noise it made called to mind, at last, my daytime sensations when a car passed—the sight and noise together. A car came roaring down hushed Edgerton Avenue in front of our house, stopped at the corner stop sign, and passed on shrieking as its engine shifted up the gears. What, precisely, came into the bedroom? A reflection from the car's oblong windshield. Why did it travel in two parts? The window sash split the light and cast a shadow.

Night after night I labored up the same long chain of reasoning, as night after night the thing burst into the room where I lay awake and Amy slept prettily and my loud heart thrashed and I froze.

There was a world outside my window and contiguous to it. If I was so all-fired bright, as my parents, who had patently no basis for comparison, seemed to think, why did I have to keep learning this same thing over and over? For I had learned it a summer ago, when men with jackhammers broke up Edgerton Avenue. I had watched them from the yard; the street came up in jagged slabs like floes. When I lay to nap, I listened. One restless afternoon I connected the new noise in my bedroom with the jackhammer men I had been seeing outside. I understood abruptly that these worlds met, the outside and the inside. I traveled the route in my mind: You walked downstairs from here, and outside from downstairs. "Outside," then, was conceivably just beyond my windows. It was the same world I reached by going out the front or the back door. I forced my imagination yet again over this route.

The world did not have me in mind; it had no mind. It was a coincidental collection of things and people, of items, and I myself was one such item—a child walking up the sidewalk, whom anyone could see or ignore. The things in the world did not necessarily cause my overwhelming feelings; the feelings were inside me, beneath my skin, behind my ribs, within my skull. They were even, to some extent, under my control.

I could be connected to the outer world by reason, if I chose, or I could yield to what amounted to a narrative fiction, to a tale of terror whispered to me by the blood in my ears, a show in light projected on the room's blue walls. As time passed, I learned to amuse myself in bed in the darkened room by entering the fiction deliberately and replacing it by reason deliberately.

When the low roar drew nigh and the oblong slid in the door, I threw my own switches for pleasure. It's coming after me; it's a car outside. It's after me. It's a car. It raced over the wall, lighting it blue wherever it ran; it bumped over Amy's maple headboard in a rush, paused, slithered elongate over the corner, shrank, flew my way, and vanished into itself with a wail. It was a car.

Our parents and grandparents, and all their friends, seemed insensible to their own prominent defect, their limp, coarse skin.

We children had, for instance, proper hands; our fluid, pliant fingers joined their skin. Adults had misshapen, knuckly hands loose in their skin like bones in bags; it was a wonder they could open jars. They were loose in their skins all over, except at the wrists and ankles, like rabbits.

We were whole, we were pleasing to ourselves. Our crystalline eyes shone from firm, smooth sockets; we spoke in pure, piping voices through dark, tidy lips. Adults were coming apart, but they neither noticed nor minded. My revulsion was rude, so I hid it. Besides, we could never rise to the absolute figural splendor they alone could on occasion achieve. Our beauty was a mere absence of decrepitude; their beauty, when they had it, was not passive but earned; it was grandeur; it was a party to power, and to artifice, even, and to knowledge. Our beauty was, in the long run, merely elfin. We could not, finally, discount the fact that in some sense they owned us, and they owned the world.

Mother let me play with one of her hands. She laid it flat on a living-room end table beside her chair. I picked up a transverse pinch of skin over the knuckle of her index finger and let it drop. The pinch didn't snap back; it lay dead across her knuckle in a yellowish ridge. I poked it; it slid over intact.

I left it there as an experiment and shifted to another finger. Mother was reading *Time* magazine.

Carefully, lifting it by the tip, I raised her middle finger an inch and released it. It snapped back to the tabletop. Her insides, at least, were alive. I tried all the fingers. They all worked. Some I could lift higher than others.

"That's getting boring."

"Sorry, Mama."

I refashioned the ridge on her index-finger knuckle; I made the ridge as long as I could, using both my hands. Moving quickly, I made parallel ridges on her other fingers—a real mountain chain, the Alleghenies; Indians crept along just below the ridgetops, eyeing the frozen lakes below them through the trees.

Skin was earth; it was soil. I could see, even on my own skin, the joined trapezoids of dust specks God had wetted and stuck with his spit the morning he made Adam from dirt. Now, all these generations later, we people could still see on our skin the inherited prints of the dust specks of Eden.

I loved this thought, and repeated it for myself often. I don't know where I got it; my parents cited Adam and Eve only in jokes. Someday I would count the trapezoids, with the aid of a mirror, and learn precisely how many dust specks Adam comprised—one single handful God wetted, shaped, blew into, and set firmly into motion and left to wander about in the fabulous garden bewildered.

The skin on my mother's face was smooth, fair, and tender; it took impressions readily. She napped on her side on the couch. Her face skin pooled on the low side; it piled up in the low corners of her deep-set eyes and drew down her lips and cheeks. How flexible was it? I pushed at a puddle of it by her nose.

She stirred and opened her eyes. I jumped back.

She reminded me not to touch her face while she was sleeping. Anybody's face.

When she sat up, her cheek and brow bone bore a deep

red gash, the mark of a cushion's welting. It was textured inside precisely with the upholstery's weave and brocade.

Another day, after a similar nap, I spoke up about this gash. I told her she had a mark on her face where she'd been sleeping.

"Do I?" she said; she ran her fingers through her hair. Her hair was short, blond, and wavy. She wore it swept back from her high, curved forehead. The skin on her forehead was both tight and soft. It would only barely shift when I tried to move it. She went to the kitchen. She was not interested in the hideous mark on her face. "It'll go away," I said. "What?" she called.

I noticed the hair on my father's arms and legs; each hair sprang from a dark dot on his skin. I lifted a hair and studied the puckered tepee of skin it pulled with it. Those hairs were in there tight. The greater the strain I put on the hair, the more puckered the tepee became, and shrunken within, concave. I could point it every which way.

"Ouch! Enough of that."

"Sorry, Daddy."

At the beach I felt my parent's shinbones. The bones were flat and curved, like the slats in a Venetian blind. The long edges were sharp as swords. But they had unexplained and, I thought, possibly diseased irregularities: nicks, bumps, small hard balls, shallow ridges, and soft spots. I was lying between my parents on an enormous towel through which I could feel the hot sand.

Loose under their shinbones, as in a hammock, hung the relaxed flesh of their calves. You could push and swing this like a baby in a sling. Their heels were dry and hard, sharp at the curved edge. The bottoms of their toes had flattened, holding the imprint of life's smooth floors even when they were lying down. I would not let this happen to me. Under certain conditions, the long bones of their feet showed under their skin. The bones rose up long and miserably thin in skeletal rays on the slopes of their feet. This terrible sight they ignored also.

In fact, they were young. Mother was twenty-two when I was born, and Father twenty-nine; both appeared to other adults much younger than they were. They were a handsome couple. I felt it overwhelmingly when they dressed for occasions. I never lost a wondering awe at the transformation of an everyday, tender, nap-creased mother into an exalted and dazzling beauty who chatted with me as she dressed.

Her blue eyes shone and caught the light, and so did the platinum waves in her hair and the pearls at her ears and throat. She was wearing a black dress. The smooth skin on her breastbone rent my heart, it was so familiar and beloved; the black silk bodice and the simple necklace set off its human fineness. Mother was perhaps a bit vain of her long and perfect legs, but not too vain for me; despite her excited pleasure, she did not share my view of her beauty.

"Look at your father," she said. We were all in the dressing room. I found him in one of the long mirrors, where he waggled his outthrust chin over the last push of his tie knot. For me he made his big ears jiggle on his skull. It was a wonder he could ever hear anything; his head was loose inside him.

Father's enormousness was an everyday, stunning fact; he was taller than everyone else. He was neither thin nor stout; his torso was supple, his long legs nimble. Before the dressing-room mirror he produced an anticipatory soft-shoe, and checked to see that his cuffs stayed down.

Now they were off. I hoped they knocked them dead; I hoped their friends knew how witty they were, and how splendid. Their parties at home did not seem very entertaining, although they laughed loudly and often fetched the one-man percussion band from the basement, or an old trumpet, or a snare drum. We children could have shown them how to have a better time. Kick the Can, for instance, never palled. A private game called Spider Cow, played by the Spencer children, also had possibilities: The spider cow hid and flung a wet washcloth at whoever found it, and erupted from hiding and chased him running all over the house.

But implicitly and emphatically, my parents and their friends were not interested. They never ran. They did not choose to run. It went with being old, apparently, and having their skin half off.

THERE WAS A BIG SNOW that same year, 1950. Traffic vanished; in the first week, nothing could move. The mailman couldn't get to us; the milkman couldn't come. Our long-legged father walked four miles with my sled to the dairy across Fifth Avenue and carried back milk.

We had a puppy, who was shorter than the big snow. Our parents tossed it for fun in the yard and it disappeared, only to pop up somewhere else at random like a loon in a lake. After a few days of this game, the happy puppy went crazy and died. It had distemper. While it was crazy it ran around the house crying, upstairs and down.

One night during the second week of the big snow I saw Jo Ann Sheehy skating on the street. I remembered this sight for its beauty and strangeness.

I was aware of the Sheehy family; they were Irish Catholics from a steep part of the neighborhood. One summer when I was walking around the block, I had to walk past skinny Tommy Sheehy and his fat father, who were hunched on their porch doing nothing. Tommy's eleven-year-old sister, Jo Ann, brought them iced tea.

"Go tell your maid she's a nigger," Tommy Sheehy said to me.

What?

He repeated it, and I did it, later, when I got home. That night, Mother came into our room after Amy was asleep. She explained, and made sure I understood. She was steely. Where had my regular mother gone? Did she hate me? She told me a passel of other words that some people use for other people. I was never to use such words, and never to associate with people who did so long as I lived; I was to apologize to Margaret Butler first thing in the morning; and I was to have no further dealings with the Sheehys.

The night Jo Ann Sheehy skated on the street, it was dark inside our house. We were having dinner in the dining room—my mother, my father, my sister Amy, who was two, and I. There were lighted ivory candles on the table. The only other light inside was the blue fluorescent lamp over the fish tank, on a sideboard. Inside the tank, neon tetras, black mollies, and angelfish circled, illumined, through the light-shot water. When I turned the fluorescent lamp off, I had learned, the fish still circled their tank in the dark. The still water in the tank's center barely stirred.

Now we sat in the dark dining room, hushed. The big snow outside, the big snow on the roof, silenced our words and the scrape of our forks and our chairs. The dog was gone, the world outside was dangerously cold, and the big snow held the houses down and the people in.

Behind me, tall chilled windows gave out onto the narrow front yard and the street. A motion must have caught my mother's eye; she rose and moved to the windows, and Father and I followed. There we saw the young girl, the transfigured Jo Ann Sheehy, skating alone under the streetlight.

She was turning on ice skates inside the streetlight's yellow cone of light—illumined and silent. She tilted and spun. She wore a short skirt, as if Edgerton Avenue's asphalt had been the ice of an Olympic arena. She wore mittens and a red knitted cap below which her black hair lifted when she turned. Under her skates the street's packed snow shone; it illumined her from below, the cold light striking her under her chin.

I stood at the tall window, barely reaching the sill; the glass fogged before my face, so I had to keep moving or hold my breath. What was she doing out there? Was everything beautiful so bold? I expected a car to run over her at any moment: the open street was a fatal place, where I was forbidden to set foot.

Once, the skater left the light. She winged into the blackness beyond the streetlight and sped down the street; only her white skates showed, and the white snow. She emerged again under another streetlight, in the continuing silence, just at our corner stop sign where the trucks' brakes hissed. Inside that second cone of light she circled backward and leaning. Then she reversed herself in an abrupt half-turn—as if she had skated backward into herself, absorbed her own motion's impetus, and rebounded from it; she shot forward into the dark street and appeared again becalmed in the first streetlight's cone. I exhaled; I looked up. Distant over the street, the night sky was moonless and foreign, a frail, bottomless black, and the cold stars speckled it without moving.

This was for many years the center of the maze, this still, frozen evening inside, the family's watching through glass the Irish girl skate outside on the street. Here were beauty and mystery outside the house, and peace and safety within. I watched passive and uncomprehending, as in summer I watched Lombardy poplar leaves turn their green sides out, and then their silver sides out—watched as if the world were a screen on which played interesting scenes for my pleasure. But there was danger in this radiant sight, in the long glimpse of the lone girl skating, for it was night, and killingly cold. The open street was fatal and forbidden. And the apparently invulnerable girl was Jo Ann Sheehy, Tommy Sheehy's sister, part of the Sheehy family, whose dark ways were a danger and a crime.

"Tell your maid she's a nigger," he had said, and when I said to Margaret, "You're a nigger," I had put myself in danger—I felt at the time, for Mother was so enraged—of being put out, tossed out in the cold, where I would go crazy and die like the dog.

That night Jo Ann alone outside in the cold had performed recklessly. My parents did not disapprove; they loved the beauty of it, and the queerness of skating on a street. The next morning I saw from the dining-room windows the street shrunken again and ordinary, tracked by tires, and the street-lights inconspicuous, and Jo Ann Sheehy walking to school in a blue plaid skirt.

Jo Ann Sheehy and the Catholic schoolchildren carried brown-and-tan workbooks, which they filled, I knew, with gibberish they not only had to memorize, they had to believe.

Every morning they filed into the subterranean maw of St. Bede's, the low stone school attached to the high stone church just a block up Edgerton Avenue. From other Protestant children, I gathered St. Bede's was a cave where Catholic children had to go to fill their brown-and-tan workbooks in the dark, possibly kneeling; they wrote down whatever the Pope said. (Whatever the Pope said, I thought, it was no prize; it didn't work; our Protestant lives were much sunnier, without our half trying.) Every afternoon, authorities "let out" the surviving children to return to their lightless steep houses, where they knelt before writhing crucifixes, bandied racial epithets about, and ate stewed fish.

One afternoon the following spring, I was sitting stilled on the side-yard swing; I was watching transparent circles swim in the sky. When I focused on them, the circles parted, as fish flew from a finger poked in their tank. Apparently it was my eyes, and not the sky, that produced the transparent circles, each with a dimple or nucleus, but I always failed to find any in my eyes in a mirror; I had tried the night before.

Now St. Bede's was, as the expression had it, letting out; Jo Ann Sheehy would walk by again, and all the other Catholic children, and perhaps the nuns. I kept an eye out for the nuns.

From my swing seat I saw the girls appear in bunches. There came Jo Ann Sheehy up the dry sidewalk with two other girls; her black hair fell over her blue blazer's back. Behind them, running back and forth across the street, little boys were throwing gravel bits. The boys held their workbooks tightly. Probably, if they lost them, they would be put to death.

In the leafy distance up Edgerton I could see a black phalanx. It blocked the sidewalk; it rolled footlessly forward like a tank. The nuns were coming. They had no bodies, and imitation faces. I quitted the swing and banged through the back door and ran in to Mother in the kitchen.

I didn't know the nuns taught the children; the Catholic children certainly avoided them on the streets, almost as much as I did. The nuns seemed to be kept in St. Bede's as in a prison, where their faces had rotted away—or they lived eyeless in the dark by choice, like bats. Parts of them were manufactured. Other parts were made of mushrooms.

In the kitchen, Mother said it was time I got over this. She took me by the hand and hauled me back outside; we crossed the street and caught up with the nuns. "Excuse me," Mother said to the black phalanx. It wheeled around. "Would you just please say hello to my daughter here? If you could just let her see your faces."

I saw the white, conical billboards they had as mock-up heads; I couldn't avoid seeing them, those white boards like pillories with circles cut out and some bunched human flesh pressed like raw pie crust into the holes. Like mushrooms and engines, they didn't have hands. There was only that disconnected saucerful of whitened human flesh at their tops. The rest, concealed by a chassis of soft cloth over hard cloth, was cylinders, drive shafts, clean wiring, and wheels.

"Why, hello," some of the top parts said distinctly. They teetered toward me. I was delivered to my enemies, and had no place to hide; I could only wail for my young life so unpityingly snuffed.

THESE ARE THE FEW, floating scenes from early childhood, from before time and understanding pinned events down to the fixed and coherent world. Soon the remembered scenes would grow in vividness and depth, as like any child I elaborated a picture of the place, and as my feelings met actual people, and as the interesting things of the world engaged my loose mind like a gear, and set it in forward motion.

A young child knows Mother as a smelled skin, a halo of light, a strength in the arms, a voice that trembles with feeling. Later the child wakes and discovers this mother—and adds facts to impressions, and historical understanding to facts.

When she was in her twenties, my mother's taste ran to modernism. In our living room on Edgerton Avenue we had a free-form blond coffee table, Jean Arp style, shaped something like a kidney, and also something like a boomerang. Over a heat register Mother hung a black iron Calder-like mobile. The mobile's disks spun and orbited slowly before a window all winter when the heat was on, and replaced for me the ensorcerizing waving of tree leaves. On the wall above the couch she hung a large print of Gauguin's *Fatata te miti;* those enormous rounded women, with their muscular curving backs, sat before a blue river in a flat and speckled jungle. On an end table she placed the first piece of art she ever bought: a Yoruba wood sculpture, a long-headed abstract woman with pointy breasts and a cold coil of wire around her neck.

Mother must have cut a paradoxical figure in her modernist living room, with her platinum blond hair, her brisk motions, her slender, urbane frame, her ironic wit (one might even say "lip")—and her wee Scotticisms. "Sit you doon," Mother said cordially to guests. If the room was too bright, she asked one of us to douse the glim. When we were babies, she bade each of us in turn, "Put your wee headie down." If no one could locate Amy when she was avoiding her nap, it was because she'd found herself a hidey-hole. Sometimes after school we discovered in our rooms a wee giftie. If Mother wanted a favor, she asked, heartrendingly, "Would you grant me a boon?"

This was all the more remarkable because Mother was no more Scotch, nor Scotch-Irish, than the Pope. She was, if anyone cared to inquire, Pennsylvania Dutch and French. But the Pittsburgh in which we lived—and that Pittsburgh only—was so strongly Scotch-Irish it might have been seventeenth-century Donegal; almost all old Pittsburgh families were Scotch-Irish. Scotticisms fairly flew in the air. And Mother picked up every sort of quaint expression.

She delighted in using queer nouns from the mountains, too. Her family hailed from Somerset, the mountain-county seat near Pittsburgh: Whiskey Rebellion country. They were pretty well educated, but they heard plenty of mountain terms.

"Where's the woolly brush?" "I need a gummy"—that is, a gum band, or rubber band. She keenly enjoyed these archaisms, and whenever she used one, she stopped enthusiastically in midsentence to list the others: "And do you know what a poke is?" We did indeed.

Her speech was an endlessly interesting, swerving path of old punch lines, heartfelt cris de coeur, puns new and old, dramatic true confessions, challenges, witty one-liners, wee Scotticisms, tag lines from Frank Sinatra songs, obsolete mountain nouns, and moral exhortations.

"I'll show him," she'd say. "I'll show him which way the bear went through the buckwheat. It'll be Katy-bar-the-door around here." "He'll be gone," Father would add wistfully, "where the woodbine twineth."

Mother woke Amy and me in the mornings by dashing into our room, wrenching aside the window curtains, cranking open our old leaded windows, shouting mysteriously, "It smells like a French whorehouse in here," and dashing out. When we got downstairs we might find her—that same morning—sitting half asleep, crumpled-of-skin in her soft bathrobe, staring at her foot in its slipper, or even with her eyes closed. If we began to whisper, we soon heard her murmur affectionately if unconvincingly into her bathrobe collar, "I'm awake."

She moved vigorously, laughed easily, spoke rapidly and boldly, and analyzed with restless force. Her moods shifted; her utterances changed key and pitch. She was fond of ending any long explanation with the sudden, puzzling kicker, "And that's why I can't imitate four Hawaiians." She stroked our heads tenderly, called us each a dozen endearing names; she thrilled, apparently, to tales of our adventures, and admired inordinately our drawings and forts. She taught us to curtsy; she taught us to play poker.

Mother's Somerset family were respectable Millers and good-looking, prominent, wild Lamberts. The Lambert women were beautiful; they married rich men. The Lambert men were charmers; they drank hard and came to early ends. They flourished during Prohibition, and set a dashing, doomed tone for the town.

Mother's handsome father was the mayor. He was so well liked that no one in town voted for his opponent. He won a contest by writing the slogan: "When better automobiles are built, Buick will build them." He and a friend journeyed to Detroit to pick up the contest prize. The trip was a famous spree; it lasted a month. He died not long after, at forty-one, when Mother was seven, and left her forever full of longing.

Late at night on Christmas Eve, she carried us each to our high bedroom, and darkened the room, and opened the window, and held us awed in the freezing stillness, saying—and we could hear the edge of tears in her voice—"Do you hear them? Do you hear the bells, the little bells, on Santa's sleigh?" We marveled and drowsed, smelling the piercingly cold night and the sweetness of Mother's warm neck, hearing in her voice so much pent emotion, feeling the familiar strength in the crook of her arms, and looking out over the silent streetlights and the chilled stars over the rooftops of the town. "Very faint, and far away—can you hear them coming?" And we could hear them coming, very faint and far away, the bells on the flying sleigh.

NEXT TO ONE OF OUR SIDE YARDS ran a short, dirty dead-end alley. We couldn't see the alley from the house; our parents had planted a row of Lombardy poplars to keep it out of sight. I found an old dime there.

High above the darkest part of the alley, in a teetering set of rooms, lived a terrible old man and a terrible old woman, brother and sister.

Doc Hall appeared only high against the sky, just outside his door at the top of two rickety flights of zigzag stairs. There he stood, grimy with coal dust, in a black suit wrinkled as underwear, and yelled unintelligibly, furiously, down at us children who played on his woodpile. We looked fearfully overhead and saw him stamp his aerial porch, a raven messing up his pile of sticks and littering the ground below. We couldn't understand his curses, but we scattered.

Doc Hall's grim sister went to early Mass at St. Bede's; she passed our house every morning. She was shapeless and sooty, dressed in black; she leaned squeezing a black cane, and walked downcast. No one knew what Mass might be; my parents shuddered to think. She crawled back and down the alley.

The alley ended at an empty, padlocked garage. In summer a few hairs of grass grew down the alley's center. Down the alley's side, broken glass, old nails, and pellets of foil and candy wrappers spiked the greasy black soil out of which a dirty catalpa and a dirty sycamore grew.

When I found the dime I was crouched in the alley digging dirt with a Popsicle stick under one of the Lombardy poplars. I struck the dime and dug around it; it was buried on edge. I pulled it out, cleaned it between my fingers, and pocketed it. Later I showed it to my father, who had been until then my only imaginable source of income. He read the date—1919—and told me it was an old dime, which might be worth more than ten cents.

He explained that the passage of time had buried the dime; soil tends to pile up around things. In Rome, he went on—looking out the kitchen window as I leaned against a counter looking up at him—in Rome, he had seen old doorways two or three stories underground. Where children had once tumbled directly outside from their doors, now visitors had to climb two flights of stairs to meet the light of the street. I stopped listening for a minute. I imagined that if the Roman children had, by awful chance, sat still in their doorways long enough, sat dreaming and forgetting to move, they, too, would have been buried in dirt, up to their chins, over their heads!—only by then, of course, they would be very old. Which was, in fact—the picture swept over me—precisely what had happened to all those Roman children, whether they sat still or not.

I turned the warm dime in my fingers. Father told me that, in general, the older a coin was, the greater its value. The older coins were farther down. I decided to devote my life to unearthing treasure. Beneath my 1919 dime, buried in the little Pittsburgh alley, might be coins older still, coins deeper down, coins from ancient times, from forgotten peoples and times, gold coins, even—pieces of eight, doubloons.

I continually imagined these old, deeply buried coins, and dreamed of them; the alley was thick with them. After I'd unearthed all the layers of wealth I could reach with a Popsicle stick, I would switch to a spade and delve down to the good stuff: to the shining layers of antique Spanish gold, of Roman gold—maybe brass-bound chests of it, maybe diamonds and rubies, maybe dulled gold from days so long past that people didn't manufacture coins at all, but simply carried bags of raw gold or ore in lumps.

That's all. It was the long years of these same few thoughts that wore tracks in my interior life. These things were mine, I figured, because I knew where to look. Because I was willing. Treasure was something you found in the alley. Treasure was something you dug up out of the dirt in a chaotic, half-forbidden, forsaken place far removed from the ordinary comings and goings of people who earned salaries in the light: under some rickety back stairs, near a falling-down pile of discarded lumber, with people yelling at you to get away from there. That I never found another old coin in that particular alley didn't matter at all.

I WALKED. My mother had given me the freedom of the streets as soon as I could say our telephone number. I walked and memorized the neighborhood. I made a mental map and located myself upon it. At night in bed I rehearsed the small world's scheme and set challenges: Find the store using backyards only. Imagine a route from the school to my friend's house. I mastered chunks of town in one direction only; I ignored the other direction, toward the Catholic church.

On a bicycle I traveled over the known world's edge, and the ground held. I was seven. I had fallen in love with a red-haired fourth-grade boy named Walter Milligan. He was tough, Catholic, from an iffy neighborhood. Two blocks beyond our school was a field—Miss Frick's field, behind Henry Clay Frick's mansion—where boys played football. I parked my bike on the sidelines and watched Walter Milligan play. As he ran up and down the length of the field, following the football, I ran up and down the sidelines, following him. After the game I rode my bike home, delirious. It was the closest we had been, and the farthest I had traveled from home.

(My love lasted two years and occasioned a bit of talk. I knew it angered him. We spoke only once. I caught him between classes in the school's crowded hall and said, "I'm sorry." He looked away, apparently enraged; his pale freckled skin flushed. He jammed his fists in his pockets, looked down, looked at me for a second, looked away, and brought out gently, "That's okay." That was the whole of it: beginning, middle, and end.)

Across the street from Walter Milligan's football field

was Frick Park. Frick Park was 380 acres of woods in residential Pittsburgh. Only one trail crossed it; the gravelly walk gave way to dirt and led down a forested ravine to a damp streambed. If you followed the streambed all day you would find yourself in a distant part of town reached ordinarily by a long streetcar ride. Near Frick Park's restful entrance, old men and women from other neighborhoods were lawn bowling on the bowling green. The rest of the park was wild woods.

My father forbade me to go to Frick Park. He said bums lived there under bridges; they had been hanging around unnoticed since the Depression. My father was away all day; my mother said I could go to Frick Park if I never mentioned it.

I roamed Frick Park for many years. Our family moved from house to house, but we never moved so far I couldn't walk to Frick Park. I watched the men and women lawn bowling—so careful the players, so dull the game. After I got a bird book I found, in the deep woods, a downy woodpecker working a tree trunk; the woodpecker looked like a jackhammer man banging Edgerton Avenue to bits. I saw sparrows, robins, cardinals, juncos, chipmunks, squirrels, and—always disappointingly, emerging from their magnificent ruckus in the leaves—pedigreed dachshunds, which a woman across the street bred.

I never met anyone in the woods except the woman who walked her shiny dachshunds there, but I was cautious, and hoped I was braving danger. If a bum came after me I would disarm him with courtesy ("Good afternoon"). I would sneak him good food from home; we would bake potatoes together under his bridge; he would introduce me to his fellow bums; we would all feed the squirrels.

The deepest ravine, over which loomed the Forbes Avenue bridge, was called Fern Hollow. There in winter I searched for panther tracks in snow. In summer and fall I imagined the woods extending infinitely. I was the first human being to see these shadowed trees, this land; I would make my pioneer clearing here, near the water. Mine would be one of those famously steep farms: "How'd you get so beat up?" "Fell out of my cornfield." In spring I pried flat rocks from the damp streambed and captured red and black salamanders. I brought the salamanders home in a bag once and terrified my mother with them by mistake, when she was on the phone.

In the fall I walked to collect buckeyes from lawns. Buckeyes were wealth. A ripe buckeye husk splits. It reveals the shining brown sphere inside only partially, as an eyelid only partially discloses an eye's sphere. The nut so revealed looks like the calm brown eye of a buck, apparently. It was odd to imagine the settlers who named it having seen more male deer's eyes in the forest than nuts on a lawn.

Walking was my project before reading. The text I read was the town; the book I made up was a map. First I had walked across one of our side yards to the blackened alley with its buried dime. Now I walked to piano lessons, four long blocks north of school and three zigzag blocks into an Irish neighborhood near Thomas Boulevard.

I pushed at my map's edges. Alone at night I added newly memorized streets and blocks to old streets and blocks, and imagined connecting them on foot. From my parents' earliest injunctions I felt that my life depended on keeping it all straight—remembering where on earth I lived, that is, in relation to where I had walked. It was dead reckoning. On darkening evenings I came home exultant, secretive, often from some exotic leafy curb a mile beyond what I had known at lunch, where I had peered up at the street sign, hugging the cold pole, and fixed the intersection in my mind. What joy, what relief, eased me as I pushed open the heavy front door!—joy and relief because, from the very trackless waste, I had located home, family, and the dinner table once again.

An infant watches her hands and feels them move. Gradually she fixes her own boundaries at the complex incurved rim of her skin. Later she touches one palm to another and tries for a game to distinguish each hand's sensation of feeling and being felt. What is a house but a bigger skin, and a neighborhood map but the world's skin ever expanding?

SOME BOYS TAUGHT ME to play football. This was fine sport. You thought up a new strategy for every play and whispered it to the others. You went out for a pass, fooling everyone. Best, you got to throw yourself mightily at someone's running legs. Either you brought him down or you hit the ground flat out on your chin, with your arms empty before you. It was all or nothing. If you hesitated in fear, you would miss and get hurt: you would take a hard fall while the kid got away, or you would get kicked in the face while the kid got away. But if you flung yourself wholeheartedly at the back of his knees—if you gathered and joined body and soul and pointed them diving fearlessly—then you likely wouldn't get hurt, and you'd stop the ball. Your fate, and your team's score, depended on your concentration and courage. Nothing girls did could compare with it.

Boys welcomed me at baseball, too, for I had, through enthusiastic practice, what was weirdly known as a boy's arm. In winter, in the snow, there was neither baseball nor football, so the boys and I threw snowballs at passing cars. I got in trouble throwing snowballs, and have seldom been happier since.

On one weekday morning after Christmas, six inches of new snow had just fallen. We were standing up to our boot tops in snow on a front yard on trafficked Reynolds Street, waiting for cars. The cars traveled Reynolds Street slowly and evenly; they were targets all but wrapped in red ribbons, cream puffs. We couldn't miss.

I was seven; the boys were eight, nine, and ten. The oldest

two Fahey boys were there—Mikey and Peter—polite blond boys who lived near me on Lloyd Street, and who already had four brothers and sisters. My parents approved Mikey and Peter Fahey. Chickie McBride was there, a tough kid, and Billy Paul and Mackie Kean too, from across Reynolds, where the boys grew up dark and furious, grew up skinny, knowing, and skilled. We had all drifted from our houses that morning looking for action, and had found it here on Reynolds Street.

It was cloudy but cold. The cars' tires laid behind them on the snowy street a complex trail of beige chunks like crenellated castle walls. I had stepped on some earlier; they squeaked. We could have wished for more traffic. When a car came, we all popped it one. In the intervals between cars we reverted to the natural solitude of children.

I started making an iceball—a perfect iceball, from perfectly white snow, perfectly spherical, and squeezed perfectly translucent so no snow remained all the way through. (The Fahey boys and I considered it unfair actually to throw an iceball at somebody, but it had been known to happen.)

I had just embarked on the iceball project when we heard tire chains come clanking from afar. A black Buick was moving toward us down the street. We all spread out, banged together some regular snowballs, took aim, and, when the Buick drew nigh, fired.

A soft snowball hit the driver's windshield right before the driver's face. It made a smashed star with a hump in the middle.

Often, of course, we hit our target, but this time, the only time in all of life, the car pulled over and stopped. Its wide black door opened; a man got out of it, running. He didn't even close the car door.

He ran after us, and we ran away from him, up the snowy Reynolds sidewalk. At the corner, I looked back; incredibly, he was still after us. He was in city clothes: a suit and tie, street shoes. Any normal adult would have quit, having sprung us into flight and made his point. This man was gaining on us. He was a thin man, all action. All of a sudden, we were running for our lives.

Wordless, we split up. We were on our turf; we could lose ourselves in the neighborhood backyards, everyone for himself. I paused and considered. Everyone had vanished except Mikey Fahey, who was just rounding the corner of a yellow brick house. Poor Mikey, I trailed him. The driver of the Buick sensibly picked the two of us to follow. The man apparently had all day.

He chased Mikey and me around the yellow house and up a backyard path we knew by heart: under a low tree, up a bank, through a hedge, down some snowy steps, and across the grocery store's delivery driveway. We smashed through a gap in another hedge, entered a scruffy backyard and ran around its back porch and tight between houses to Edgerton Avenue; we ran across Edgerton to an alley and up our own sliding woodpile to the Halls' front yard; he kept coming. We ran up Lloyd Street and wound through mazy backyards toward the steep hilltop at Willard and Lang.

He chased us silently, block after block. He chased us silently over picket fences, through thorny hedges, between houses, around garbage cans, and across streets. Every time I glanced back, choking for breath, I expected he would have quit. He must have been as breathless as we were. His jacket strained over his body. It was an immense discovery, pounding into my hot head with every sliding, joyous step, that this ordinary adult evidently knew what I thought only children who trained at football knew: that you have to fling yourself at what you're doing, you have to point yourself, forget yourself, aim, dive.

Mikey and I had nowhere to go, in our own neighborhood or out of it, but away from this man who was chasing us. He impelled us forward; we compelled him to follow our route. The air was cold; every breath tore my throat. We kept running, block after block; we kept improvising, backyard after backyard, running a frantic course and choosing it simultaneously, failing always to find small places or hard places to slow him down, and discovering always, exhilarated, dismayed, that only bare speed could save us—for he would never give up, this man—and we were losing speed.

He chased us through the backyard labyrinths of ten

blocks before he caught us by our jackets. He caught us and we all stopped.

We three stood staggering, half blinded, coughing, in an obscure hilltop backyard: a man in his twenties, a boy, a girl. He had released our jackets, our pursuer, our captor, our hero: he knew we weren't going anywhere. We all played by the rules. Mikey and I unzipped our jackets. I pulled off my sopping mittens. Our tracks multiplied in the backyard's new snow. We had been breaking new snow all morning. We didn't look at each other. I was cherishing my excitement. The man's lower pants legs were wet; his cuffs were full of snow, and there was a prow of snow beneath them on his shoes and socks. Some trees bordered the little flat backyard, some messy winter trees. There was no one around: a clearing in a grove, and we the only players.

It was a long time before he could speak. I had some difficulty at first recalling why we were there. My lips felt swollen; I couldn't see out of the sides of my eyes; I kept coughing.

"You stupid kids," he began perfunctorily.

We listened perfunctorily indeed, if we listened at all, for the chewing out was redundant, a mere formality, and beside the point. The point was that he had chased us passionately without giving up, and so he had caught us. Now he came down to earth. I wanted the glory to last forever.

But how could the glory have lasted forever? We could have run through every backyard in North America until we got to Panama. But when he trapped us at the lip of the Panama Canal, what precisely could he have done to prolong the drama of the chase and cap its glory? I brooded about this for the next few years. He could only have fried Mikey Fahey and me in boiling oil, say, or dismembered us piecemeal, or staked us to anthills. None of which I really wanted, and none of which any adult was likely to do, even in the spirit of fun. He could only chew us out there in the Panamanian jungle, after months or years of exalting pursuit. He could only begin, "You stupid kids," and continue in his ordinary Pittsburgh accent with his normal righteous anger and the usual common sense.

If in that snowy backyard the driver of the black Buick had cut off our heads, Mikey's and mine, I would have died happy, for nothing has required so much of me since as being chased all over Pittsburgh in the middle of winter—running terrified, exhausted—by this sainted, skinny, furious red-headed man who wished to have a word with us. I don't know how he found his way back to his car.

Our parents would sooner have left us out of Christmas than leave us out of a joke. They explained a joke to us while they were still laughing at it; they tore a still-kicking joke apart, so we could see how it worked. When we got the first Tom Lehrer album in 1954, Mother went through the album with me, cut by cut, explaining. B.V.D.s are men's underwear. Radiation makes you sterile, and lead protects from radiation, so the joke is . . .

Our father kept in his breast pocket a little black notebook. There he noted jokes he wanted to remember. Remembering jokes was a moral obligation. People who said, "I can never remember jokes," were like people who said, obliviously, "I can never remember names," or "I don't bathe."

"No one tells jokes like your father," Mother said. Telling a good joke well—successfully, perfectly—was the highest art. It was an art because it was up to you: if you did not get the laugh, you had told it wrong. Work on it, and do better next time. It would have been reprehensible to blame the joke, or, worse, the audience.

As we children got older, our parents discussed with us every technical, theoretical, and moral aspect of the art. We tinkered with a joke's narrative structure: "Maybe you should begin with the Indians." We polished the wording. There is a Julia Randall story set in Baltimore which we smoothed together for years. How does the lady word the question? Does she say, "How are you called?" No, that is needlessly awkward. She just says, "What's your name?" And he says, "Folks generally call me Bominitious." No, he can just say, "They call me Bominitious."

We analyzed many kinds of pacing. We admired with Father the leisurely meanders of the shaggy-dog story. "A young couple moved to the Swiss Alps," one story of his began, "with their grand piano"; and ended, to a blizzard of thrown napkins, ". . . Oppernockity tunes but once." "Frog goes into a bank," another story began, to my enduring pleasure. The joke was not great, but with what a sweet light splash you could launch it! "Frog goes into a bank," you said, and your canoe had slipped delicately and surely into the water, into Lake Champlain with painted Indians behind every tree, and there was no turning back.

Father was also very fond of stories set in bars that starred zoo animals or insects. These creatures apparently came into bars all over America, either accompanied or alone, and sat down to face incredulous, sarcastic bartenders. (It was a wonder the bartenders were always so surprised to see talking dogs or drinking monkeys or performing ants, so surprised year after year, when clearly this sort of thing was the very essence of bar life.) In the years he had been loose, swinging aloft in the airy interval between college and marriage, Father had frequented bars in New York, listening to jazz. Bars had no place whatever in the small Pittsburgh world he had grown up in, and lived in now. Bars were so far from our experience that I had assumed, in my detective work, that their customers were ipso facto crooks. Father's bar jokes—"and there were the regulars, all sitting around"—gave him the raffish air of a man who was at home anywhere. (How poignant were his "you knows" directed at me: you know how bartenders are; you know how the regulars would all be sitting around. For either I, a nine-year-old girl, knew what he was talking about, then or ever, or nobody did. Only because I read a lot, I often knew.)

Our mother favored a staccato, stand-up style; if our father could perorate, she could condense. Fellow goes to a psychiatrist. "You're crazy." "I want a second opinion!" "You're ugly." "How do you get an elephant out of the theater? You can't; it's in his blood."

What else in life so required, and so rewarded, such care?

"Tell the girls the one about the four-by-twos, Frank."

"Let's see. Let's see."

"Fellow goes into a lumberyard . . ."

"Yes, but it's tricky. It's a matter of point of view." And Father would leave the dining room, rubbing his face in concentration, or as if he were smearing on greasepaint, and return when he was ready.

"Ready with the four-by-twos?" Mother said.

Our father hung his hands in his pockets and regarded the far ceiling with fond reminiscence.

"Fellow comes into a lumberyard," he began.

"Says to the guy, 'I need some four-by-twos.' 'You mean two-by-fours?' 'Just a minute. I'll find out.' He walks out to the parking lot, where his buddies are waiting in the car. They roll down the car window. He confers with them awhile and comes back across the parking lot and says to the lumberyard guy, 'Yes. I mean two-by-fours.'

"Lumberyard guy says, 'How long do you want them?' 'Just a minute,' fellow says, 'I'll find out.' He goes out across the parking lot and confers with the people in the car and comes back across the parking lot to the lumberyard and says to the guy, 'A long time. We're building a house.' "

After any performance Father rubbed the top of his face with both hands, as if it had all been a dream. He sat back down at the dining-room table, laughing and shaking his head. "And when you tell a joke," Mother said to Amy and me, "laugh. It's mean not to."

We were brought up on the classics. Our parents told us all the great old American jokes, practically by number. They collaborated on, and for our benefit specialized in, the painstaking paleontological reconstruction of vanished jokes from extant tag lines. They could vivify old *New Yorker* cartoons, source of many tag lines. The lines themselves—"Back to the old drawing board," and "I say it's spinach and I say the hell with it," and "A simple yes or no will suffice"—were no longer funny; they were instead something better, they were fixtures in the language. The tag lines of old jokes were the most powerful expressions we learned at our parents' knees. A few words suggested a complete story and a wealth of feelings. Learning our culture backward, Amy and Molly and I heard only later about *The Divine Comedy* and the Sistine Chapel ceiling, and still later about the Greek and Roman myths, which held no residue of feeling for us at all—certainly not the vibrant suggestiveness of old American jokes and cartoons.

Our parents reserved a few select jokes, such as "Archibald a Soulbroke," like vintage wines for extraordinary occasions. We heard about or witnessed those rare moments—maybe three or four in a lifetime—when circumstances combined to float our father to the top of the world, from which precarious eminence he would consent to fling himself into "Archibald a Soulbroke."

Telling "Archibald a Soulbroke" was for Father an exhilarating ordeal, like walking a tightrope over Niagara Falls. It was a long, absurdly funny, excruciatingly tricky tour de force he had to tell fast, and it required beat-perfect concentration. He had to go off alone and rouse himself to an exalted, superhuman pitch in order to pace the hot coals of its dazzling verbal surface. Often enough he returned from his prayers to a crowd whose moment had passed. We knew that when we were grown, the heavy, honorable mantle of this heart-pounding joke would fall on us.

There was another very complicated joke, also in a select category, which required a long weekend with tolerant friends.

You had to tell a joke that was not funny. It was a long, pointless story about a construction job that ended with someone's throwing away a brick. There was nothing funny about it at all, and when your friends did not laugh, you had to pretend you'd muffed it. (Your husband in the crowd could shill for you: " 'Tain't funny, Pam. You told it all wrong.")

A few days later, if you could contrive another occasion for joke telling, and if your friends still permitted you to speak, you set forth on another joke, this one an old nineteenth-century chestnut about angry passengers on a train. The lady plucks the lighted, smelly cigar from the man's mouth and flings it from the moving train's window. The man seizes the little black poodle from her lap and hurls the poor dog from the same window. When at last the passengers draw unspeaking into the station, what do they see coming down the platform but the black poodle, and guess what it has in its mouth? "The cigar," say your friends, bored sick and vowing never to spend another weekend with you. "No," you say, triumphant, "the brick." This was Mother's kind of joke. Its very riskiness excited her. It wasn't funny, but it was interesting to set up, and it elicited from her friends a grudging admiration.

How long, I wondered, could you stretch this out? How boldly could you push an audience—not, in Mother's terms, to "slay them," but to please them in some grand way? How could you convince the listeners that you knew what you were doing, that the payoff would come? Or conversely, how long could you lead them to think you were stupid, a dumb blonde, to enhance their surprise at the punch line, and heighten their pleasure in the good story you had controlled all along? Alone, energetic and trying to fall asleep, or walking the residential streets long distances every day, I pondered these things.

Our parents were both sympathetic to what professional comedians call flop sweat. Boldness was all at our house, and of course you would lose some. Anyone could be misled by poor judgment into telling a "woulda hadda been there." Telling a funny story was harder than telling a joke; it was trying out, as a tidy unit, some raveling shred of the day's fabric. You learned to gauge what sorts of thing would "tell." You learned that some people, notably your parents, could rescue some things by careful narration from the category "woulda hadda been there" to the category "it tells."

At the heart of originating a funny story was recognizing it as it floated by. You scooped the potentially solid tale from the flux of history. Once I overheard my parents arguing over a thirty-year-old story's credit line. "It was my mother who said that," Mother said. "Yes, but"—Father was downright smug—"I was the one who noticed she said that."

The sight gag was a noble form, and the running gag was a noble form. In combination they produced the top of the line, the running sight gag, like the sincere and deadpan Nairobi Trio interludes on Ernie Kovacs. How splendid it was when my parents could get a running sight gag going. We heard about these legendary occasions with a thrill of family pride, as other children hear about their progenitors' war exploits.

The sight gag could blur with the practical joke—not a noble form but a friendly one, which helps the years pass. My parents favored practical jokes of the sort you set up and then retire from, much as one writes books, possibly because imagining people's reactions beats witnessing them. They procured a living hen and "hypnotized" it by setting it on the sink before the bathroom mirror in a friend's cottage by the New Jersey shore. They spent weeks constructing a ten-foot sea monster—from truck inner tubes, cement blocks, broomsticks, lumber, pillows—and set it afloat in a friend's pond. On Sanibel Island, Florida, they baffled the shell collectors each Saint Patrick's Day by boiling a bucketful of fine shells in green dye and strewing the green shells up and down the beach before dawn. I woke one Christmas morning to find in my stocking, hung from the mantel with care, a leg. Mother had charmed a department store display manager into lending her one.

When I visited my friends, I was well advised to rise when their parents entered the room. When my friends visited me, they were well advised to duck.

Central in the orders of merit, and the very bread and butter of everyday life, was the crack. Our mother excelled at the crack. We learned early to feed her lines just to watch her speed to the draw. If someone else fired a crack simultaneously, we compared their concision and pointedness and declared a winner.

Feeding our mother lines, we were training as straight men. The straight man's was an honorable calling, a bit like that of the rodeo clown: despised by the ignorant masses, perhaps, but revered among experts who understood the skills required and the risks run. We children mastered the deliberate misunderstanding, the planted pun, the Gracie

Allen know-nothing remark, which can make of any interlocutor an instant hero.

How very gracious is the straight man!—or, in this case, the straight girl. She spreads before her friend a gift-wrapped, beribboned gag line he can claim for his own, if only he will pick it up instead of pausing to contemplate what a nitwit he's talking to.

OUR FATHER'S PARENTS LIVED IN PITTSBURGH; Amy and I dined with them, rather formally, every Friday night until dancing school swept us away. Our grandfather's name was, like our father's, Frank Doak. He was a banker, a potbellied, bald man with thin legs: a generous-hearted, joking, calm Pittsburgher of undistinguished Scotch-Irish descent, who held his peace. Our grandmother's name was Meta Waltenburger Doak. We children called her Oma, accenting both syllables. She was an imperious and kindhearted grande dame of execrable taste, a tall, potbellied redhead, the proud descendant and heir of well-to-do Germans in Louisville, Kentucky, who boasted that she never worked a day in her life. Our father was their only child.

Every summer these grandparents moved to their summer house on the shore of Lake Erie, near North Madison, Ohio, and every summer Amy and I moved in with them for a month or two. With them also lived Mary Burinda, a thin woman who still carried a buzzing trace of Hungarian at the tip of her tongue, and who cooked and cleaned and warmly befriended both our grandmother and us; and Henry Watson, a Pittsburgh man who drove the car, tended the grounds, and served dinner.

Oma was odd about money. One ordinary summer afternoon at Lake Erie, I found a penny in the sand.

"Money!" Oma said. "If you've found money, don't touch it with your bare hands. You don't know who has touched it."

My bare hands? Oma, Amy, and I had been swimming

at the beach below the house when I found the penny. Now I was to bring it to Oma for safekeeping, and go wash my hands in the Lake as well as I could. This washing ought to hold calamity at bay until we could get to the bathhouse to take showers.

Oma had told me that when she was in her teens, she had sewed rows of lace on her chemises, to bring her bust forward. It was hard to believe. By the time I knew her, her bust was enormous. Walking beside Amy and me up the path to the bathhouse, she cut an imposing figure: her legs were long and fine, her hips slender, her carriage erect. She wore her red hair short, in waves. Her face was round; her head was round and slightly flattened vertically, like Raggedy Ann's. Her blue eyes were small, stubby-lashed; her nose was short and bulbous. The expression on her thin lips was sometimes peevish, sometimes doting.

In the bathhouse Amy and I peeled down our bathing suits. Stuck to my belly skin, as if by suction, were flat bits of big Lake Erie sand—gray and smooth, like hammered dots. I pried them off with a fingernail. My buttocks were cold, my arms hot.

We all stood in the women's shower; we stamped our sandy feet on the shower's cedar-slat floor, and turned on the water. Oma soaped her soft arms with the red sponge. When it was my turn to use the red sponge, I got sand in it. I washed myself down with soap and sand—a delicate operation on sunburned shoulders, a pleasingly rough one on poison-ivy-covered shins.

Peering cheerfully down at me through the sharp strands of water, Oma said, "Have you washed your hands very well with soap?" She stuck her round head under the nozzle, screwed her eyes tight shut, and wagged her chin.

I mistook bodies for persons, and admired Oma above all for her freckles. Also, she could float. She could float on her back in Lake Erie, she said, and read a book. Sadly, I never saw her perform this feat, for she was not so much of a reader that she felt the need of reading while bathing, but I often saw her float for long periods. Her vast tight abdomen rose in the air; her fingers joined over it. She could easily have held a book. Her small round head in its white rubber cap lay half submerged. From the shore I could see an expression of benignity or complacency on her features, features which had been rather bunched together, centered around her nose, by the tight bathing cap and its strap. She rocked over the little waves, calm as a plank. She wore white tennis shoes into the water, for our part of Lake Erie was bumpy with glacial stones. When she floated, her tennis shoes stuck straight up.

From the bathhouse we climbed two flights of stairs to the house proper, a mid-twenties white frame house with five bedrooms and three bathrooms upstairs, and more on the third floor, where Henry Watson lived.

Now Henry was pushing a mower over the back lawn. Politely he asked us how the water was; he didn't like the water.

Henry rarely wore his full uniform at the Lake; he wore only the heavy black pants, a white shirt, and suspenders. When he drove, he put on his cap. Famously, Henry loved summers at the Lake. He took pride in the cool lawns with their bluish, cylindrical grass. Mornings he cleared the horsetail beside the long path from the bathhouse. He washed the glass porch walls. He stood in the driveway up to his ankles in foam, a ridged black garden hose in his hand, washing the car. Vapor rose low from the hot asphalt driveway; it was warm in the nostrils, sweet, smelling of soft soap. Henry's gold-rimmed glasses flashed.

In Pittsburgh, during the rest of the year, Henry went home every night to the Homewood section. By day he waited at curbs while my grandmother tried on shoes. He served dinner, nightly, in his white uniform jacket. Here at the Lake he had one friend, another chauffeur, named Cicero. He slept on the third floor. On a kitchen counter was his drinking glass.

Inside, Oma and Amy and I found Mary Burinda standing on the back of a flower-print couch. She held against a living-room window a curtain rod from which depended heavy, flower-print curtains. "Here, Mrs. Doak? Or lower?" Our grandfather was watching the Cleveland Indians on television in the same room. Henry would join him when he finished mowing.

"No, higher, I should think. But not now."

Mary climbed from the couch. She was thin, sallow-skinned, full of love, quick to laugh. She always wore her white uniform. By choice, she rarely came to the beach. I asked her how long it would be until dinner. She looked at her black watch. Two hours, she said. You *kids*. How was the water?

Mary was forty-five, to Oma's sixty-five. She had lived with them twenty-four years. Almost all of her family, she told me, had died one day during the 1918 flu epidemic; her parents and most of her brothers and sisters had died one after the other in the house. Both at the Lake and in Pittsburgh she had a room and a private bath; over the bed hung a crucifix, the most bizarre object I had ever seen. Of Mary's Catholicism, Oma used to say, with a tinge of admiration, "She's stubborn."

Mary and Henry ate in the kitchen. We ate on the enclosed porch. From the porch we could see the tall fir trunks on the back lawn, and the lake below and far down the cliff, the lake beating in waves over the stones and up onto the sand, and blurring offshore with the sky.

Oma settled in for a phone call. She combed her wet hair and shaped its waves with a freckled forefinger. She sat to her Florentine leather desk, by the tall living-room windows. I joined my grandfather at the Cleveland Indians game; Amy rolled around bored on the floor. I could hear Oma. She placed the call with the operator and apparently got a busy signal, for she hung up, called the operator, and shouted that she'd like to try again. There was a silence. Then she lost her temper. "But I just *told* you. I'm calling Marie Phillips in Pittsburgh—I just this minute finished telling you. Have you already *lost* the number I gave you?"

Oma had grown up an only child, in some luxury. There was something Victorian about her. Her grasp of the great world was slender. She believed that there was not only a telephone operator assigned to her, but also a burglar. She and my grandfather had Cadillacs, one at a time. She referred to the car as "the machine": "Henry is coming around with the machine."

At the Lake, Oma wore cotton sundresses and low-heeled sandals. She relaxed there; we all did. She barely resembled the formidable woman she was in Pittsburgh the rest of the year. In Pittsburgh, she dressed. She wore jewelry by the breastful, by the armload: diamonds, rubies, emeralds. She wore big rings like engine bearings, and vast, slithering mink coats. She wore purple and green silk, purple and green linen, purple and green wool—dresses, suits, robes—and leather high-heeled pumps, which drew attention to her long, energetic legs and thin ankles. She looked imposing. She looked, we at our house tended to think—for how females looked occupied most of females' attention—terrible. We were all blondes; we disliked purple, we disliked green, and were against the rest of it, too.

American Standard Corporation started as a plumbing brass foundry in Louisville, Kentucky. Oma's grandfather, Theodore Ahrens, came over from Hamburg, Germany, in 1848 and opened that foundry, which kept growing. The family kept holdings in the firm. Our grandmother was not ashamed that she was German. Amy and I were ashamed of being one-fourth German because of her (never guessing that our own mother, whose hatred of things German was an ordinary part of family politics, was in fact half German herself).

I thought Oma was brilliant to have accepted the suit of Frank Doak. He was an uncommonly kind and good-natured man. Oma had met him in 1914, while she was visiting Pittsburgh cousins. He was from an ordinary Scotch-Irish family so devotedly Presbyterian they forbade looking at the Sunday funnies. (William Doak had immigrated from Ulster in 1848 with a cargo of woolens. He wrote home depressed that the socks weren't selling well. The name Doak was a corruption of McDougal.) Oma had been a spoiled, fun-loving, red-haired beauty; our grandfather handled her with the same solid calm that is reputedly so effective on racehorses.

By the time I knew him, our grandfather was a vice-president of Pittsburgh's Fidelity Trust Bank. He looked very like a cartoonist's version of "vested interests." In fact, he almost always wore a vest, and a gold watch on a chain; he was short and heavy; he had a small white mustache; he smoked cigars. At home, his thin legs crossed under his belly, he read the financial section of the paper, tolerant of children who might have been driven, in the long course of waiting for dinner, to beating their fingertips on his scalp.

From almost every room at the Lake house, you could see Lake Erie and its mild shore. From my bed as soon as I woke, I gauged the waves' height: two inches, three. The waves disintegrated on the big beach; from the high cliff where our house stood, their breaking sounded like poured raw rice. By afternoon, the waves were two or three feet high. They seemed to rattle the glass porch windows; they broke on the long beaches like seas. On the horizon we saw ore boats—lakers—bringing iron ore east from the Masabi ore range near Lake Superior. Ships had been carrying iron ore bound for Pittsburgh across Lake Erie since the time of Carnegie and Frick. Sometimes a dusting of ore washed up or blew up on the sand beach. It lay in scalloped windrows, as did the powdery purple garnet grains after storms.

Canada, we knew, lay across the Lake. Many times I planned to run away to Canada; I would lie on the canvas raft and paddle with my hands. Instead I took up bicycle exploring. I rode a bicycle all morning for months, for years. I saw apple orchards, nurseries, and cornfields.

The land I toured mornings on a bike was flat and fertile. The Ohio settlers had a crazy way of clearing this land of forest. Father told me about it one night after dinner (our parents visited the Lake every summer). The pioneers, he said—the Scotch-Irish, German, and English pioneers—came in and sawed halfway through the trunk of every tree they wanted to fell, every tree in—was I to believe this?—several acres. Then when a wind came up they felled some big trees at the forest's upwind edge, and those trees took the whole forest down, just knocked those half-cut trees before them like dominoes. I laughed—what a good idea. Father laughed. When you saw through a tree trunk, he said, the first half is a lot easier than the second half. They never had to saw through the second halves.

I rode past cantaloupe stands and truck farms planted in tomatoes. I rode past sandy woods and frame houses with green shutters and screened porches full of kids. I played baseball with some of the kids. I got a book on birds, took up bird-watching, and saw a Baltimore oriole in an apple orchard. I straddled my bike in amazement, bare feet on the cool morning road, and watched the brilliant thing bounce singing from treetop to treetop in the sun.

I learned to whistle; I whistled "The Wayward Wind." I sang "The Wayward Wind," too, at the top of my lungs for an hour one evening, bored on the porch, hurling myself from chair to chair singing, and wondering when these indulgent grandparents would stop me. At length my grandfather looked up from his paper and said, "That's a sad song you're singing. Do you know that?" And I was amazed he knew that. Did he yearn to wander, my banker grandfather, like the man in "The Wayward Wind"?

Afternoons we swam at our own beach. When Grandfather joined us I stared at the skin on his legs. It consisted of many scaly layers of fragile translucency, which together appeared bluish. On it, white starbursts appeared at random, and red streaks were visible somewhere inside. The stars and stripes forever. The skin on Oma's legs was similarly translucent; the freckles seemed to float flat just below the first few layers.

I found a beachful of neighborhood kids to swim with; I came home only to eat. Evenings Amy and I played cards with Oma and Mary on the porch, or, when we were younger, we colored in coloring books with Oma. Oma was a tidy hand with a crayon. She fought with us over the crayons as an equal. The big woodland silk moths banged at the glass walls beside our bare shoulders under the lamp.

We left the Lake by rising at three, eating the last of the sweet cantaloupe by lamplight, and driving through horse-and-buggy Mennonite country back to Pittsburgh. We retraced one of the routes the old Indian traders had used in the 1750s, back from the Lake Erie country to the Forks of the Ohio, where they could load up on trinkets and, pretty soon, buy a drink. In Pittsburgh, Oma would go back to work. Although she claimed never to have worked, in fact she and a partner directed the Presbyterian Hospital gift shop as volunteers full time for twenty years. And in Pittsburgh this year, Amy and I would start new schools.

Now in the embarrassing Cadillac we pulled up in front of our house. From the capacious row of jump seats Amy and I were delivered—suntanned, cheerful, covered with poison ivy, and in possession of suitcases full of new green and purple dresses—to our mother.

The rivalry between our mother and Oma was intense; it was a long, civilized antagonism. Our mother had won the moral battle—we children were shamed, for instance, by Oma's bursts of bigotry—but Mother fought on for autonomy, seeking to prevent our being annexed to Oma's big tribe of Louisville Germans. When I was a baby, Oma had several times hauled me downriver to Louisville for Christmas as a prize; Mother put a stop to it.

If Oma had a great deal of shockingly loose money, we had, we fancied, good taste. Oma had a green-and-blue blown-glass sculpture of two intertwined swans, full of bubbles; we had a black iron Calder-style mobile. Oma had a servant and a companion. We had help. Our "help" shared our drinking glasses. At our parents' parties, friends ate lasagna and danced; at our grandparents' parties, guests ate sauerbraten and went to the theater.

Matters of taste are not, it turns out, moral issues. We thought we were grander than Oma morally, that she was bigoted and vain, quite as if we ourselves were neither. Actually it was her taste we most deplored. We thought that merely possessing a gaudy figurine was a worse offense than wholeheartedly embracing snobbery. We could not see how clearly she saw us, two small children just about to start prep school, who enjoyed the fruits of her family's prosperity, and who had barely peeped beyond Pittsburgh. She never said a word against our mother. But like our mother, she never gave up the struggle, even, apparently, after she suffered a stroke—for after her stroke she earnestly asked our father from time to time, "Have you ever thought of marrying?"

He pressed her freckled hand. Of course we loved her.

It was not, in retrospect, a fair fight. For at our house, we were all so young.

WE HAD MOVED WHEN I WAS EIGHT. We moved from Edgerton Avenue to Richland Lane, a hushed dead-end street on the far side of Frick Park. We expanded into a brick house on two lots. There was a bright sunporch under buckeye trees; there was a golden sandstone wall with fireplace and bench that Mother designed, which ran the length of the living room.

It was into this comfortable house that the last of us sisters, Molly, was born, two years later. It was from this house that Father would leave to go down the river to New Orleans, and to this house that he would return early, from the river at Louisville. Here Mother told the contractor where she wanted kitchen walls knocked out. Here on the sunporch Amy tended her many potentially well-dressed dolls, all of whom were, unfortunately, always sick in bed. Here I began a life of reading books, and drawing, and playing at the sciences. Here also I began to wake in earnest, and shed superstition, and plan my days.

Every August when Amy and I returned from the Lake, we saw that workmen had altered the house in our absence—the dining room seemed bigger, the kitchen was lighter—but we couldn't recall how it had been. I thought Mother was a genius for thinking up these improvements, for the house always seemed fine to me, yet it got better and better.

This August, the summer I was ten, we returned from the Lake and found our shared room uncannily tidy and stilled, dark, while summer, the summer in which we had been immersed, played outside the closed windows like a movie.

So it always was, those first few minutes in an emptied room. They made you self-conscious; you felt yourself living your life. As soon as you unzipped your suitcase and opened the window, you broke the spell; you plunged again into the rush and weather.

While we were gone, Molly had learned to crawl. She pulled herself up and stood singing in her playpen on the flat part of the front lawn; the buckeye boughs stirred far overhead, and waved over her round arms their speckled lights.

Usually when it was hot the family swam at the distant country-club pool. Now that we were back from the Lake, all that resumed—a nasty comedown after the Lake, to whose neighborhood beach I had gone alone, and where we were all kids among kids who owned the beach and our days. There, at the Lake, if you wanted to leave, you simply kicked the bike's kickstand and sprang into the seat and away, in one skilled gesture like cowboys' mounting horses, rode away on the innocent Ohio roads under old, still trees. At the country club, you often wanted to leave as soon as you had come, but there was no leaving to be had. The country-club pool drew a society as complex and constraining, if not so entertaining, as any European capital's drawing room did. You forgot an old woman's name at some peril to your entire family. What if you actually, physically, ran into her? Knocked her off her pins? It was no place for children.

One country-club morning this August, I saw a red blotch moving in a dense hedge by the club's baby pool. I crept up on the red blotch in my cold bathing suit and discovered that it was a rose-breasted grosbeak. I had never seen one. This living, wild bird, which could fetch up anyplace it pleased, had inexplicably touched down at our country club. It scratched around in a hedge between the baby pool and the sixth hole. The dumb cluck, why a country club?

Mother said Father was going down the river in his boat pretty soon. It sounded like a swell idea.

One windy Saturday morning, after the Lake and before the new private school started, I hung around the house. It was too early for action in the neighborhood. To wake up, I read on the sunporch.

The sunporch would wake anybody up; Father had now put on the record: Sharkey Bonano, "Li'l Liza Jane." He was bopping around, snapping his fingers; now he had wandered outside and stood under the big buckeye trees. I could see him through the sunporch's glass walls. He peered up at a patch of sky as if it could tell him, old salt that he was, right there on Richland Lane, how the weather would be next week on the Ohio River.

I was starting *Kidnapped.* It began in Scotland; David Balfour's father asked that a letter be delivered "when the house is redd up." Some people in Pittsburgh redd up houses, too. The hardworking parents of my earliest neighborhood friends said it: You kids redd up this room. It meant clean up, or ready up. I never expected to find "redd up" in so grand a thing as a book. Apparently it was Scots. I hadn't heard the phrase since we moved.

I rode back to Edgerton Avenue from time to time after we moved—to look around, and to fix in my mind the route back: past the lawn bowlers in Frick Park, past the football field, and beyond the old elementary schoolyard, where a big older boy had said to me, "Why, you're a regular Ralph Kiner." Touring that old neighborhood, I saw the St. Bede's nuns. I sped past them, careless, on my bike.

"Redd up," David Balfour's father said in *Kidnapped.* I was reading on the sunporch, on the bright couch. "Oh, Li'l Liza!" said the music on the record, "Li'l Liza Jane." Next week Father was going down the river to New Orleans. Maybe they'd let him sit in a set on the drums; maybe Zutty Singleton would be there and holler out to him—"Hey, Frank!"

The wind rattled the windowed sunporch walls beside me. I could see, without getting up, some green leaves blowing down from the buckeye branches overhead. Everything in the room was bright, even the bookshelves, even Amy's melancholy dolls. The blue shadows of fast clouds ran over the far walls and floor. Father snapped his fingers and wandered, tall and loose-limbed, over the house.

I was ten years old now, up into the double numbers, where I would likely remain till I died. I am awake now forever, I thought suddenly; I have converged with myself in the present. My hands were icy from holding *Kidnapped* up; I always read lying down. I felt time in full stream, and I felt consciousness in full stream joining it, like the rivers.

Part Two

WE LIVED IN A CLEAN CITY whose center was new; after the war, a few business leaders and Democratic Mayor David L. Lawrence had begun cleaning it up. Beneath the new city, and tucked up its hilly alleys, lay the old Pittsburgh, and the old foothill land beneath it. It was all old if you dug far enough. Our Pittsburgh was like Rome, or Jericho, a palimpsest, a sliding pile of cities built ever nearer the sky, and rising ever higher over the rivers. If you dug, you found things.

Oma's chauffeur, Henry Watson, dug a hole in our yard on Edgerton Avenue to plant a maple tree when I was born, and again when Amy was born three years later. When he dug the hole for Amy's maple, he found an arrowhead—smaller than a dime and sharp. Our mother continually remodeled each of the houses we lived in: the workmen knocked out walls and found brick walls under the plaster and oak planks under the brick. City workers continually paved the streets: they poured asphalt over the streetcar tracks, streetcar tracks their fathers had wormed between the old riverworn cobblestones, cobblestones laid smack into the notorious nineteenth-century mud. Long stretches of that mud were the same pioneer roads that General John Forbes's troops had hacked over the mountains from Carlisle, or General Braddock's troops had hacked from the Chesapeake and the Susquehanna, widening with their axes the woodland paths the Indians had worn on deer trails.

Many old stone houses had slate-shingle roofs. I used to find blown shingles cracked open on the sidewalk; some of them bore—inside, where no one had been able to look until now—fine fossil prints of flat leaves. I heard there were dinosaur bones under buildings. The largest coal-bearing rock sequence in the world ran under Pittsburgh and popped out at Coal Hill, just across the Monongahela. (Then it ducked far underground and ran up into Nova Scotia, dove into the water and crossed under the Atlantic, and rolled up again thick with coal in Wales.) There were layers of natural gas beneath Pittsburgh, and pools of petroleum the pioneers called Seneca oil, because only Indians would fool with it.

We children lived and breathed our history—our Pittsburgh history, so crucial to the country's story and so typical of it as well—without knowing or believing any of it. For how can anyone know or believe stories she dreamed in her sleep, information for which and to which she feels herself to be in no way responsible? A child is asleep. Her private life unwinds inside her skin and skull; only as she sheds childhood, first one decade and then another, can she locate the actual, historical stream, see the setting of her dreaming private life—the nation, the city, the neighborhood, the house where the family lives—as an actual project under way, a project living people willed, and made well or failed, and are still making, herself among them. I breathed the air of history all unaware, and walked oblivious through its littered layers.

Outside in the neighborhoods, learning our way around the streets, we played among the enormous stone monuments of the millionaires—both those tireless Pittsburgh founders of the heavy industries from which the nation's wealth derived (they told us at school) and the industrialists' couldn't-lose bankers and backers, all of whom began as canny boys, the stories of whose rises to riches adults still considered inspirational to children.

We were unthinkingly familiar with the moguls' immense rough works as so much weird scenery on long drives. We saw the long, low-slung stripes of steel factories by the rivers; we saw pyramidal heaps of yellow sand at glassworks by the shining railroad tracks; we saw rusty slag heaps on the outlying hilltops, and coal barges tied up at the docks. We recognized, on infrequent trips downtown, the industries' smooth corporate headquarters, each to its own soaring building—Gulf Oil, Alcoa, U.S. Steel, Koppers Company, Pittsburgh Plate Glass, Mellon Bank. Our classmates' fathers worked in these buildings, or at nearby corporate headquarters for Westinghouse Electric, Jones & Laughlin Steel, Rockwell Manufacturing, American Standard, Allegheny Ludlum, Westinghouse Air Brake, and H. J. Heinz.

The nineteenth-century industrialists' institutions—galleries, universities, hospitals, churches, Carnegie libraries, the Carnegie Museum, Frick Park, Mellon Park—were, many of them, my stomping grounds. These absolute artifacts of philanthropy littered the neighborhoods with marble. Millionaires' encrusted mansions, now obsolete and turned into parks or art centers, weighed on every block. They lent their expansive, hushed moods to the Point Breeze neighborhoods where we children lived and where those fabulous men had lived also, or rather had visited at night in order to sleep. Everywhere I looked, it was the Valley of the Kings, their dynasty just ended, and their monuments intact but already out of fashion.

All these immensities wholly dominated the life of the city. So did their several peculiar social legacies: their powerful Calvinist mix of piety and acquisitiveness, which characterized the old and new Scotch-Irish families and the nation they helped found; the walled-up hush of what was, by my day, old money—amazing how fast it ages if you let it alone—and the clang and roar of making that money; the owners' Presbyterian churches, their anti-Catholicism, anti-Semitism, Republicanism, and love of continuous work; their dogmatic practicality, their easy friendliness, their Pittsburgh-centered innocence, and, paradoxically, their egalitarianism.

For all the insularity of the old guard, Pittsburgh was always an open and democratic town. "Best-natured people I ever went among," a Boston visitor noted two centuries earlier. In colonial days, everybody went to balls, regardless of rank. No one had any truck with aristocratic pretensions—hadn't they hated the British lords in Ulster? People who cared to rave about their bloodlines, Mother told us, had stayed in Europe, which deserved them. We were vaguely proud of living in a city so full of distinctive immigrant groups, among which we never thought to number ourselves. We had no occasion to visit the steep hillside neighborhoods—Polish, Hungarian, Rumanian, Italian, Slav—of the turn-of-the-century immigrants who poured the steel and stirred the glass and shoveled the coal.

We children played around the moguls' enormous pale stone houses, restful as tombs, houses set back just so on their shaded grounds. Henry Clay Frick's daughter, unthinkably old, lived alone in her proud, sinking mansion; she had lived alone all her life. No one saw her. Men mowed the wide lawns and seeded them, and pushed rollers over them, over the new grass seed and musket balls and arrowheads, over the big trees' roots, bones, shale, coal.

We knew bits of this story, and we knew none of it. Odd facts stuck in the mind: On the Pennsylvania frontier in the eighteenth century, people pressed hummingbirds as if they were poppies, between pages of heavy books, and mailed them back to Ulster and Scotland as curiosities. Money was so scarce in the western Pennsylvania mountains that, as late as the mid-nineteenth century, people substituted odds and ends like road contracts, feathers, and elderberries.

We knew that before big industry there had been small industry here—H. J. Heinz setting up a roadside stand to sell horseradish roots from his garden. There were the makers of cannonballs for the Civil War. There were the braggart and rowdy flatboat men and keelboat men, and the honored steamboat builders and pilots. There were local men getting rich in iron and glass manufacturing and trade downriver. There was a whole continentful of people passing through, native-born and immigrant men and women who funneled down Pittsburgh, where two rivers converged to make a third river. It was the gateway to the West; they piled onto flatboats and launched out into the Ohio River singing, to head for new country. There had been a Revolutionary War, and before that the French and Indian War. And before that, and first of all, had been those first settlers come walking bright-eyed in, into nowhere from out of nowhere, the people who, as they said, "broke wilderness," the pioneers. This was the history.

I treasured some bits; they provided doll-like figures for imagination's travels and wars. There in private imagination were the vivid figures of history in costume, tricked out as if for amateur outdoor drama: a moving, clumsy, insignificant spectacle like everything else the imagination proposes to itself for pure pleasure only—nothing real, nobody gets hurt, it's only ketchup.

WHILE FATHER WAS MOTORING down the river, my reading was giving me a turn.

At a neighbor boy's house, I ran into Kimon Nicolaides' *The Natural Way to Draw*. This was a manual for students who couldn't get to Nicolaides' own classes at New York's Art Students League. I was amazed that there were books about things one actually did. I had been drawing in earnest, but at random, for two years. Like all children, when I drew I tried to reproduce schema. The idea of drawing from life had astounded me two years previously, but I had gradually let it slip, and my drawing, such as it was, had sunk back into facile sloth. Now this book would ignite my fervor for conscious drawing, and bind my attention to both the vigor and the detail of the actual world.

For the rest of August, and all fall, this urgent, hortatory book ran my life. I tried to follow its schedules: every day, sixty-five gesture drawings, fifteen memory drawings, an hour-long contour drawing, and "The Sustained Study in Crayon, Clothed" or "The Sustained Study in Crayon, Nude."

While Father was gone, I outfitted an attic bedroom as a studio, and moved in. Every summer or weekend morning at eight o'clock I taped that day's drawing schedule to a wall. Since there was no model, nude or clothed, I drew my baseball mitt.

I drew my baseball mitt's gesture—its tense repose, its expectancy, which ran up its hollows like a hand. I drew its contours—its flat fingertips strung on square rawhide thongs. I drew its billion grades of light and dark in detail, so the glove weighed vivid and complex on the page, and the trapezoids small as dust motes in the leather fingers cast shadows, and the pale palm leather was smooth as a belly and thick. "Draw anything," said the book. "Learning to draw is really a matter of learning to see," said the book. "Imagine that your pencil point is touching the model instead of the paper." "All the student need concern himself with is reality."

With my pencil point I crawled over the mitt's topology. I slithered over each dip and rise; I checked my bearings, admired the enormous view, and recorded it like Meriwether Lewis mapping the Rockies.

One thing struck me as odd and interesting. A gesture drawing took forty-five seconds; a Sustained Study took all morning. From any still-life arrangement or model's pose, the artist could produce either a short study or a long one. Evidently, a given object took no particular amount of time to draw; instead the artist took the time, or didn't take it, at pleasure. And, similarly, things themselves possessed no fixed and intrinsic amount of interest; instead things were interesting as long as you had attention to give them. How long does it take to draw a baseball mitt? As much time as you care to give it. Not an infinite amount of time, but more time than you first imagined. For many days, so long as you want to keep drawing that mitt, and studying that mitt, there will always be a new and finer layer of distinctions to draw out and lay in. Your attention discovers—seems thereby to produce—an array of interesting features in any object, like a lamp.

By noon, all this drawing would have gone to my head. I slipped into the mitt, quit the attic, quit the house, and headed up the street, looking for a ball game.

My friend had sought permission from his father for me to borrow *The Natural Way to Draw;* it was his book. Grown men and growing children rarely mingled then. I had lived two doors away from this family for several years, and had never clapped eyes on my good friend's father; still, I now regarded him as a man after my own heart. Had he another book about drawing? He had; he owned a book about pencil drawing. This book began well enough, with the drawing of trees. Then it devoted a chapter to the schematic representation of shrubbery. At last it dwindled into its true subject, the drawing of buildings.

My friend's father was an architect. All his other books were about buildings. He had been a boy who liked to draw, according to my friend, so he became an architect. Children who drew, I learned, became architects; I had thought they became painters. My friend explained that it was not proper to become a painter; it couldn't be done. I resigned myself to architecture school and a long life of drawing buildings. It was a pity, for I disliked buildings, considering them only a stiffer and more ample form of clothing, and no more important.

I began reading books, reading books to delirium. I began by vanishing from the known world into the passive abyss of reading, but soon found myself engaged with surprising vigor because the things in the books, or even the things surrounding the books, roused me from my stupor. From the nearest library I learned every sort of surprising thing—some of it, though not much of it, from the books themselves.

The Homewood branch of Pittsburgh's Carnegie Library system was in a Negro section of town—Homewood. This branch was our nearest library; Mother drove me to it every two weeks for many years, until I could drive myself. I only very rarely saw other white people there.

I understood that our maid, Margaret Butler, had friends in Homewood. I never saw her there, but I did see Henry Watson.

I was getting out of Mother's car in front of the library when Henry appeared on the sidewalk; he was walking with some other old men. I had never before seen him at large; it must have been his day off. He had gold-rimmed glasses, a gold front tooth, and a frank, open expression. It would embarrass him, I thought, if I said hello to him in front of his friends. I was wrong. He spied me, picked me up—books and all—swung me as he always did, and introduced Mother and me to his friends. Later, as we were climbing the long stone steps to the library's door, Mother said, "That's what I mean by good manners."

The Homewood Library had graven across its enormous stone facade: FREE TO THE PEOPLE. In the evenings, neighborhood people—the men and women of Homewood—browsed in the library, and brought their children. By day, the two vaulted rooms, the adults' and children's sections, were almost empty. The kind Homewood librarians, after a trial period, had given me a card to the adult section. This was an enormous silent room with marble floors. Nonfiction was on the left.

Beside the farthest wall, and under leaded windows set ten feet from the floor, so that no human being could ever see anything from them—next to the wall, and at the farthest remove from the idle librarians at their curved wooden counter, and from the oak bench where my mother waited in her camel's-hair coat chatting with the librarians or reading—stood the last and darkest and most obscure of the tall nonfiction stacks: NEGRO HISTORY and NATURAL HISTORY. It was in Natural History, in the cool darkness of a bottom shelf, that I found *The Field Book of Ponds and Streams*.

The Field Book of Ponds and Streams was a small, blue-bound book printed in fine type on thin paper, like *The Book of Common Prayer*. Its third chapter explained how to make sweep nets, plankton nets, glass-bottomed buckets, and killing jars. It specified how to mount slides, how to label insects on their pins, and how to set up a freshwater aquarium.

One was to go into "the field" wearing hip boots and perhaps a head net for mosquitoes. One carried in a "rucksack" half a dozen corked test tubes, a smattering of screw-top baby-food jars, a white enamel tray, assorted pipettes and eyedroppers, an artillery of cheesecloth nets, a notebook, a hand lens, perhaps a map, and *The Field Book of Ponds and Streams*. This field—unlike the fields I had seen, such as the field where Walter Milligan played football—was evidently very well watered, for there one could find, and distinguish among, daphniae, planaria, water pennies, stonefly larvae, dragonfly nymphs, salamander larvae, tadpoles, snakes, and turtles, all of which one could carry home.

That anyone had lived the fine life described in Chapter 3 astonished me. Although the title page indicated quite plainly that one Ann Haven Morgan had written *The Field Book of Ponds and Streams,* I nevertheless imagined, perhaps from the authority and freedom of it, that its author was a man. It would be good to write him and assure him that someone had found his book, in the dark near the marble floor at the Homewood Library. I would, in the same letter or in a subsequent one, ask him a question outside the scope of his book, which was where I personally might find a pond, or a stream. But I did not know how to address such a letter, of course, or how to learn if he was still alive.

I was afraid, too, that my letter would disappoint him by betraying my ignorance, which was just beginning to attract my own notice. What, for example, was this noisome-sounding substance called cheesecloth, and what do scientists do with it? What, when you really got down to it, was enamel? If candy could, notoriously, "eat through enamel," why would anyone make trays out of it? Where—short of robbing a museum—might a fifth-grade student at the Ellis School on Fifth Avenue obtain such a legendary item as a wooden bucket?

The Field Book of Ponds and Streams was a shocker from beginning to end. The greatest shock came at the end.

When you checked out a book from the Homewood Library, the librarian wrote your number on the book's card and stamped the due date on a sheet glued to the book's last page. When I checked out *The Field Book of Ponds and Streams* for the second time, I noticed the book's card. It was almost full. There were numbers on both sides. My hearty author and I were not alone in the world, after all. With us, and sharing our enthusiasm for dragonfly larvae and single-celled plants, were, apparently, many Negro adults.

Who were these people? Had they, in Pittsburgh's Homewood section, found ponds? Had they found streams? At home, I read the book again; I studied the drawings; I reread Chapter 3; then I settled in to study the due-date slip. People read this book in every season. Seven or eight people were reading this book every year, even during the war.

Every year, I read again *The Field Book of Ponds and Streams*. Often, when I was in the library, I simply visited it. I sat on the marble floor and studied the book's card. There we all were. There was my number. There was the number of someone else who had checked it out more than once. Might I contact this person and cheer him up? For I assumed that, like me, he had found pickings pretty slim in Pittsburgh.

The people of Homewood, some of whom lived in visible poverty, on crowded streets among burned-out houses—they dreamed of ponds and streams. They were saving to buy microscopes. In their bedrooms they fashioned plankton nets. But their hopes were even more vain than mine, for I was a child, and anything might happen; they were adults, living in Homewood. There was neither pond nor stream on the streetcar routes. The Homewood residents whom I knew had little money and little free time. The marble floor was beginning to chill me. It was not fair.

I had been driven into nonfiction against my wishes. I wanted to read fiction, but I had learned to be cautious about it.

"When you open a book," the sentimental library posters said, "anything can happen." This was so. A book of fiction was a bomb. It was a land mine you wanted to go off. You wanted it to blow your whole day. Unfortunately, hundreds of thousands of books were duds. They had been rusting out of everyone's way for so long that they no longer worked. There was no way to distinguish the duds from the live mines except to throw yourself at them headlong, one by one.

The suggestions of adults were uncertain and incoherent. They gave you Nancy Drew with one hand and *Little Women* with the other. They mixed good and bad books together because they could not distinguish between them. Any book which contained children, or short adults, or animals, was felt to be a children's book. So also was any book about the sea—as though danger or even fresh air were a child's prerogative—or any book by Charles Dickens or Mark Twain.

Virtually all British books, actually, were children's books; no one understood children like the British. Suited to female children were love stories set in any century but this one. Consequently one had read, exasperated often to fury, *Pickwick Papers, Désirée, Wuthering Heights, Lad, a Dog, Gulliver's Travels, Gone With the Wind, Robinson Crusoe,* Nordhoff and Hall's *Bounty* trilogy, *Moby-Dick, The Five Little Peppers, Innocents Abroad, Lord Jim, Old Yeller.*

The fiction stacks at the Homewood Library, their volumes alphabetized by author, baffled me. How could I learn to choose a novel? That I could not easily reach the top two shelves helped limit choices a little. Still, on the lower shelves I saw too many books: Mary Johnson, *Sweet Rocket;* Samuel Johnson, *Rasselas;* James Jones, *From Here to Eternity.* I checked out the last because I had heard of it; it was good. I decided to check out books I had heard of. I had heard of *The Mill on the Floss.* I read it, and it was good. On its binding was printed a figure, a man dancing or running; I had noticed this figure before. Like so many children before and after me, I learned to seek out this logo, the Modern Library colophon.

The going was always rocky. I couldn't count on Modern Library the way I could count on, say, *Mad* magazine, which never failed to slay me. *Native Son* was good, *Walden* was pretty good, *The Interpretation of Dreams* was okay, and *The Education of Henry Adams* was awful. *Ulysses,* a very famous book, was also awful. *Confessions* by Augustine, whose title promised so much, was a bust. *Confessions* by Jean-Jacques Rousseau was much better, though it fell apart halfway through.

In fact, it was a plain truth that most books fell apart halfway through. They fell apart as their protagonists quit, without any apparent reluctance, like idiots diving voluntarily into buckets, the most interesting part of their lives, and entered upon decades of unrelieved tedium. I was forewarned, and would not so bobble my adult life; when things got dull, I would go to sea.

Jude the Obscure was the type case. It started out so well. Halfway through, its author forgot how to write. After Jude got married, his life was over, but the book went on for hundreds of pages while he stewed in his own juices. The same thing happened in *The Little Shepherd of Kingdom Come,* which Mother brought me from a fair. It was simply a hazard of reading. Only a heartsick loyalty to the protagonists of the early chapters, to the eager children they had been, kept me reading chronological narratives to their bitter ends. Perhaps later, when I had become an architect, I would enjoy the latter halves of books more.

This was the most private and obscure part of life, this Homewood Library: a vaulted marble edifice in a mostly decent Negro neighborhood, the silent stacks of which I plundered in deep concentration for many years. There seemed then, happily, to be an infinitude of books.

I no more expected anyone else on earth to have read a book I had read than I expected someone else to have twirled the same blade of grass. I would never meet those Homewood people who were borrowing *The Field Book of Ponds and Streams;* the people who read my favorite books were invisible or in hiding, underground. Father occasionally raised his big eyebrows at the title of some volume I was hurrying off with, quite as if he knew what it contained—but I thought he must know of it by hearsay, for none of it seemed to make much difference to him. Books swept me away, one after the other, this way and that; I made endless vows according to their lights, for I believed them.

THE INTERIOR LIFE EXPANDS AND FILLS; it approaches the edge of skin; it thickens with its own vivid story; it even begins to hear rumors, from beyond the horizon skin's rim, of nations and wars. You wake one day and discover your grandmother; you wake another day and notice, like any curious naturalist, the boys.

There were already boys then: not tough boys—much as I missed their inventiveness and easy democracy—but the polite boys of Richland Lane. The polite boys of Richland Lane aspired to the Presbyterian ministry. Their fathers were surgeons, lawyers, architects, and businessmen, who sat on the boards of churches and hospitals. Early on warm weekday evenings, we children played rough in the calm yards and cultivated woods, grabbing and bruising each other often enough in the course of our magnificently organized games. On Saturday afternoons, these same neighborhood boys appeared wet-combed and white-shirted at the front door, to take me gently to the movies on the bus. And there were the dancing-school boys, who materialized at the front door on Valentine's Day, holding heart-shaped boxes of chocolates.

I was ten when I met the dancing-school boys; it was that same autumn, 1955. Father was motoring down the river. The new sandstone wall was up in the living room.

Outside the city, the mountainside maples were turning; the oaks were green. Everywhere in the spreading Mississippi watershed, from the Allegheny and the Ohio here in Pittsburgh to the Missouri and the Cheyenne and the Bighorn draining the Rocky Mountains, yellow and red leaves, silver-maple and black-oak leaves, or pale cottonwood leaves and aspen, slipped down to the tight surface of the moving water. A few leaves fell on the decks of Father's boat when he tied up at an Ohio island for lunch; he raked them off with his fingers, probably, and thought it damned strange to be raking leaves at all.

Molly, the new baby, had grown less mysterious; she smiled and crawled over the grass or the rug. The family had begun spending summers around a country-club pool. Amy and I had started at a girls' day school, the Ellis School; I belted on the green jumper I would wear, in one size or another, for the next eight years, until I left Pittsburgh altogether. I was taking piano lessons, art classes. And I started dancing school.

The dancing-school boys, it turned out, were our boys, the boys, who ascended through the boys' private school as we ascended through the girls'. I was surprised to see them that first Friday afternoon in dancing school. I was surprised, that is, to see that I already knew them, that I already knew almost everyone in the room; I was surprised that dancing school, as an institution, was eerily more significant than all my other lessons and classes, and that it was not peripheral at all, but central.

For here we all were. I'd seen the boys in, of all places, church—one of the requisite Presbyterian churches of Pittsburgh. I'd seen them at the country club, too. I knew the girls from church, the country club, and school. Here we all were at dancing school; here we all were, dressed to the teeth and sitting on rows of peculiar painted and gilded chairs. Here we all were, boys and girls, plunged by our conspiring elders into the bewildering social truth that we were meant to make each other's acquaintance. Dancing school.

There in that obscure part of town, there in that muffled enormous old stone building, among those bizarre and mismatched adults who seemed grimly to dance their lives away in that dry and claustrophobic ballroom—there, it proved, was the unlikely arena where we were foreordained to assemble, Friday after Friday, for many years until the distant and seemingly unrelated country clubs took over the great work of providing music for us later and later into the night until the time came when we should all have married each other up, at last.

"Isn't he cute?" Bebe would whisper to me as we sat in the girls' row on the edge of the ballroom floor. I had never before seen a painted chair; my mother favored wood for its own sake. The lugubrious instructors were demonstrating one of several fox-trots.

Which?

"Ronny," she whispered one week, and "Danny," the next. I would find that one in the boys' row. He'd fastened his fists to his seat and was rocking back and forth from his hips all unconsciously, open-mouthed.

Sure.

"Isn't he cute?" Mimsie would ask at school, and I would think of this Ricky or Dick, recall some stray bit of bubbling laughter in which he had been caught helpless, pawing at his bangs with his bent wrist, his saliva whitening his braces' rubber bands and occasionally forming a glassy pane at the corner of his mouth; I would remember the way his head bobbed, and imagine those two parallel rods at the back of his neck, which made a thin valley where a short tip of hair lay tapered and curled; the way he scratched his ear by wincing, raising a shoulder, and rubbing the side of his head on his jacket's sleeve seam. Cute?

You bet he was cute. They all were.

Onstage the lonely pianist played "Mountain Greenery." Sometimes he played "Night and Day." It was Friday afternoon; we could have been sled riding. On Fridays, our unrelated private schools, boys' and girls', released us early. On Fridays, dancing school met, an hour later each year, until at last we met in the dark, disrupting our families' dinners, and at last certain boys began to hold our hands, carefully looking away, after a given dance, to secure us for the next one.

We all wore white cotton gloves. Only with the greatest of effort could I sometimes feel, or fancy I felt, the warmth of a boy's hand—through his glove and my glove—on my right palm. My gloved left hand lay lightly, always lightly, on his jacket shoulder. His gloved right hand lay, forgotten by both of us, across the clumsy back of my dress, across its lumpy velvet bow or its long cold zipper concealed by brocade.

Between dances when we held hands, we commonly interleaved our fingers, as if for the sheer challenge of it, for our thick cotton gloves permitted almost no movement, and we quickly cut off the circulation in each other's fingers. If for some reason we had released each other's hands quickly, without thinking, our gloves would have come off and dropped to the ballroom floor together still entwined, while our numbed bare fingers slowly regained sensation and warmth.

We were all on some list. We were to be on that list for life, it turned out, unless we left. I had no inkling of this crucial fact, although the others, I believe, did. I was mystified to see that whoever devised the list misunderstood things so. The best-liked girl in our class, my friend Ellin Hahn, was conspicuously excluded. Because she was precisely fifty percent Jewish, she had to go to Jewish dancing school. The boys courted her anyway, one after the other, and only made do with the rest of us at dancing school. From other grades at our school, all sorts of plain, unintelligent, lifeless girls were included. These were quiet or silly girls, who seemed at school to recognize their rather low places, but who were unreasonably exuberant at dancing school, and who were gradually revealed to have known all along that in the larger arena they occupied very high places indeed. And these same lumpish, plain, very rich girls wound up marrying, to my unending stupefaction, the very liveliest and handsomest of the boys.

The boys. There were, essentially, a dozen or so of them and a dozen or so of us, so it was theoretically possible, as it were, to run through all of them by the time you finished school. We saw our dancing-school boys everywhere we went. Yet they were by no means less extraordinary for being familiar. They were familiar only visually: their eyebrows we could study in quick glimpses as we danced, eyebrows that met like spliced ropes over their noses; the winsome whorls of their hair we could stare at openly in church, hair that radiated spirally from the backs of their quite individual skulls; the smooth skin on their pliant torsos at the country-club pool, all so fascinating, each so different; and their weird little graceful bathing suits: the boys. Richard, Rich, Richie, Ricky, Ronny, Donny, Dan.

They called each other witty names, like Jag-Off. They could dribble. They walked clumsily but assuredly through the world, kicking things for the hell of it. By way of conversation, they slugged each other on their interesting shoulders.

They moved in violent jerks from which we hung back, impressed and appalled, as if from horses slamming the slats of their stalls. This and, as we would have put it, their messy eyelashes. In our heartless, condescending, ignorant way we loved their eyelashes, the fascinating and dreadful way the black hairs curled and tangled. That's the kind of vitality they had, the boys, that's the kind of novelty and attraction: their very eyelashes came out amok, and unthinkably original. That we loved, that and their cloddishness, their broad, vaudevillian reactions. They were always doing slow takes. Their breathtaking lack of subtlety in every particular, we thought—and then sometimes a gleam of consciousness in their eyes, as surprising as if you'd caught a complicit wink from a brick.

Ah, the boys. How little I understood them! How little I even glimpsed who they were. How little any of us did, if I may extrapolate. How completely I condescended to them when we were ten and they were in many ways my betters. And when we were fifteen, how little I understood them still, or again. I still thought they were all alike, for all practical purposes, no longer comical beasts now but walking gods who conferred divine power with their least glances. In fact, they were neither beasts nor gods, as I should have guessed. If they were alike it was in this, that all along the boys had been in the process of becoming responsible members of an actual and moral world we small-minded and fast-talking girls had never heard of.

They had been learning self-control. We had failed to develop any selves worth controlling. We were enforcers of a code we never questioned; we were vigilantes of the trivial. They had been accumulating information about the world outside our private schools and clubs. We had failed to notice that there was such a thing. The life of Pittsburgh, say, or the United States, or assorted foreign continents, concerned us no more than Jupiter did, or its moons.

The boys must have shared our view that we were, as girls, in the long run, negligible—not any sort of factor in anybody's day, or life, no sort of creatures to be reckoned with, or even reckoned in, at all. For they could perhaps see that we possessed neither self-control nor information, so the world could not be ours.

There was something ahead of the boys, we all felt, but we didn't know what it was. To a lesser extent and vicariously, it was ahead of us, too. From the quality of attention our elders gave to various aspects of our lives, we could have inferred that we were being prepared for a life of ballroom dancing. But we knew that wasn't it. Only children practiced ballroom dancing, for which they were patently unsuited. It was something, however, that ballroom dancing obliquely prepared us for, just as, we were told, the study of Latin would obliquely prepare us for something else, also unspecified.

Whatever we needed in order to meet the future, it was located at the unthinkable juncture of Latin class and dancing school. With the declension of Latin nouns and the conjugation of Latin verbs, it had to do with our minds' functioning; presumably this held true for the five steps of the fox-trot as well. Learning these things would permanently alter the structure of our brains, whether we wanted it to or not.

So the boys, with the actual world before them, had when

they were small a bewildered air, and an endearing and bravura show of manliness. On the golden-oak ballroom floor, every darkening Friday afternoon while we girls rustled in our pastel dresses and felt at our hair ineffectually with our cotton gloves, the boys in their gloves, standing right in plain view between dances, exploded firecrackers. I would be waltzing with some arm-pumping tyke of a boy when he whispered excitedly in my ear, "Guess what I have in my pocket?" I knew. It was a cherry bomb. He slammed the thing onto the oak floor when no one was looking but a knot of his friends. The instructors flinched at the bang and stiffened; the knot of boys scattered as if shot; we could taste the sharp gunpowder in the air, and see a dab of gray ash on the floor. And when he laughed, his face reddened and gave off a vaporous heat. He seemed tickled inside his jiggling bones; he flapped his arms and slapped himself and tears fell on his tie.

They must have known, those little boys, that they would inherit corporate Pittsburgh, as indeed they have. They must have known that it was theirs by rights as boys, a real world, about which they had best start becoming informed. And they must have known, too, as Pittsburgh Presbyterian boys, that they could only just barely steal a few hours now, a few years now, to kid around, to dribble basketballs and explode firecrackers, before they were due to make a down payment on a suitable house.

Soon they would enter investment banking and take their places in the management of Fortune 500 corporations. Soon in their scant spare time they would be serving on the boards of schools, hospitals, country clubs, and churches. No wonder they laughed so hard. These were boys who wore ties from the moment their mothers could locate their necks.

I assumed that like me the boys dreamed of running away to sea, of curing cancer, of playing for the Pirates, of painting in Paris, of tramping through the Himalayas, for we were all children together. And they may well have dreamed these things, and more, then and later. I don't know.

Those boys who confided in me later, however, when we

were all older, dreamed nothing of the kind. One wanted to be top man at Gulf Oil. One wanted to accumulate a million dollars before he turned thirty. And one wanted to be majority leader in the U.S. Senate.

But these, the boys who confided in me, were the ones I would love when we were in our teens, and they were, according to my predilection, not the dancing-school boys at all, but other, oddball boys. I would give my heart to one oddball boy after another—to older boys, to prep-school boys no one knew, to him who refused to go to college, to him who was a hood, and all of them wonderfully skinny. I loved two such boys deeply, one after the other and for years on end, and forsook everything else in life, and rightly so, to begin learning with them that unplumbed intimacy that is life's chief joy. I loved them deeply, one after the other, for years on end, I say, and hoped to change their worldly ambitions and save them from the noose. But they stood firm.

And it could be, I think, that only those oddball boys, none of whom has inherited Pittsburgh at all, longed to star in the world of money and urban power; and it could be that the central boys, our boys, who are now running Pittsburgh responsibly, longed to escape. I don't know. I never knew them well enough to tell.

AMY WAS A LOOKER; I privately thought she must be the most beautiful child on earth. She inherited our father's thick, wavy hair. Her eyes were big, and so were her lashes; her nose was delicate and fluted, her skin translucent. Her mouth curved quaintly; her lips fitted appealingly, as a cutter's bow dents and curls the water under way. Plus she was quiet. And little, and tidy, and calm, and more or less obedient. She had an endearing way—it attracted even me—of standing with her legs tight together, and peering up and around with wild, stifled hilarity and parodied curiosity, as if to see if—by chance—anybody has noticed small her and found her amusing.

At the top of Richland Lane lived Amy's friend Tibby, a prematurely sophisticated blond tot, best remembered for having drawled conversationally to Mother, when she, Tibby, was only six and still missing her front teeth, "I love your hair, Mrs. Doak." When Tibby and Amy were eight, Amy brought home yet another straight-A report card. Shortly afterward, Mother overheard Tibby say exasperated to Amy, "How can you be so smart in school and so dumb after school?" In fact, as the years passed, after school became Amy's bailiwick, and she was plenty smart at it.

When Amy wasn't playing with Tibby, she played with her dolls. They were a hostile crew. Lying rigidly in their sickbeds, they shot at each other a series of haughty expletives. She had picked these up from Katy Keene comic books; Katy Keene was a society girl with a great many clothes. Amy pronounced every consonant of these expletives: Humph, pshaw.

"I'll show you, you vixen!" cried a flat-out and staring piece of buxom plastic from its Naturalizer shoe box.

"Humph!"

"Pshaw!"

"Humph!"

"Pshaw!"

We all suffered a bit for want of more of these words.

I had made several attempts to snuff baby Amy in her cradle. Mother had repeatedly discovered me pouring glasses of water carefully into her face. So when Molly had appeared, Mother had led me to believe the new baby was a kind of present for me. Actually, the baby displaced Amy. I liked everything about her—the strong purity of her cheeriness, bewilderment, outrage; her big dumb baldness, pointy fingers, little teeth, the works.

Molly possessed a dingy blanket, which she trailed behind her like a travois on her crawls. During this period, she held the belief that when she herself could not see, she was invisible. Consequently, in order to hide, she draped her head in this blanket. When it was time for her nap, we found her a pyramidal woolly mound on the pantry floor, a veritable monadnock, her fat foot protruding from the blanket's edge. She barely breathed from suspense. It broke her heart to be discovered and bundled away, day after day; she tried hard to hide ever more motionless.

When the spirit of Lister seized Mother, she flung the appalling blanket into the washing machine. Molly wept inconsolably, so Mother carried her to the basement to let her watch the thing go around in the dryer. Molly plumped down intently, straight-backed, before the dryer as if it were a television screen; her big head rolled around and around on her tiny neck. Mother, Amy, and I watched from the top of the stairs, trying not to let her hear us. Finally, Mother cut the blanket in two so she could wash one easily, and that particular joke was over.

After Father got back from his river trip, he needed something to do. He had an income, but the days themselves, if not the coffers, needed filling. So he joined as its business manager an offbeat outfit that made radio spots in its recording studio, and also rented the studio to all comers. The company was small enough so that he got to do some acting, which he loved. He practiced around the house, saying in rotund tones, for my amusement, "Hello, Horatio." The line came from a story I liked, about one of his friends' acting lessons at the American Academy in New York. The budding actors stood in the opened window over Fifty-sixth Street and intoned, over and over again, "Hello, Horatio." The idea was to say "Hello, Horatio" not loudly but deeply, in a voice so resonant that passersby far below would look up. That was the test. The window was high above the street. Did anyone look up? Then the actor had boomed his speech well. Once I was playing mumblety-peg with my friend Pin Ford on the side lawn under the buckeye trees when I heard it from my parents' upstairs window: "Hello, Horatio!" I looked right up.

Pittsburgh was a great town for radio—KDKA was the world's first commercial radio station—and a great town for KDKA's funny-voiced radio characters, like Omicron, a little fellow from outer space. Father's senior partner in the recording studio was the voice of some of these characters. Father ran the business end of the enterprise, and sales, and in those years he had a good time. The people there called him Paco. He did some straight advertising spots, and got called from his desk to help out with crowd noises—what radio people call Walla Walla talk. A mere two people, he said, could sound like a great crowd—a lively cocktail party or a muted full house at the theater—if they continuously muttered, "Pork chops and Lyonnaise potatoes."

This was not the way any other man we knew lived. Our father had been reared, for instance, cheek by jowl with Oma's best friend's son, Edgar Speer. They played together summers at Lake Erie; they spent holidays together. Our family still spent some holidays with the Speers and their boys, but now Edgar Speer—Uncle Ed—was pretty busy; he was executive vice-president of U.S. Steel, and soon would be president, and then chairman. "Edgar's, er, promotion," Oma called the last, uncomfortable.

Much later, Father and his company got involved in the making of a low-budget local horror movie, in which Father played a scientist interviewed on television. The name of the movie was *Night of the Living Dead*. It was a startling success both in Europe and in the United States. First Mother was angry that he was involved in a horror movie, and then she was angry that he hadn't got a percentage of it.

Not only that, but the Pirates were in the cellar again. They lived in the cellar, like trolls. They hadn't won a pennant since 1927. Nobody could even remember when they won ball games, the bums. They had some hitters, but no pitchers.

On the yellow back wall of our Richland Lane garage, I drew a target in red crayon. The target was a batter's strike zone. The old garage was dark inside; I turned on the bare bulb. Then I walked that famously lonely walk out to the mound, our graveled driveway, and pitched.

I squinted at the strike zone, ignoring the jeers of the batter—oddly, Ralph Kiner. I received no impressions save those inside the long aerial corridor that led to the target. I threw a red-and-blue rubber ball, one of those with a central yellow band. I wound up; I drew back. The target held my eyes. The target set me spinning as the sun from a distance winds the helpless spheres. Entranced and drawn, I swung through the moves and woke up with the ball gone. It felt as if I'd gathered my own body, pointed it carefully, and thrown it down a tunnel bored by my eyes.

I pitched in a blind fever of concentration. I pitched, as I did most things, in order to concentrate. Why do elephants drink? To forget. I loved living at my own edge, as an explorer on a ship presses to the ocean's rim; mind and skin were one joined force curved out and alert, prow and telescope. I pitched, as I did most things, in a rapture.

Now here's the pitch. I followed the ball as if it had been my own head, and watched it hit the painted plastered wall. High and outside; ball one. While I stood still stupefied by the effort of the pitch, while I stood agog, unbreathing, mystical, and unaware, here came the daggone rubber ball again, bouncing out of the garage. And I had to hustle up some snappy fielding, or lose the ball in a downhill thicket next door.

The red, blue, and yellow ball came spinning out to the driveway, and sprang awry on the gravel; if I nabbed it, it was apt to bounce out of my mitt. Sometimes I threw the fielded grounder to first—sidearm—back to the crayon target, which had become the first baseman. Fine, but the moronic first baseman spat it back out again at once, out of the dark garage and bouncing crazed on the gravel; I bolted after it, panting. The pace of this game was always out of control.

So I held the ball now, and waited, and breathed, and fixed on the target till it mesmerized me into motion. In there, strike one. Low, ball two.

Four balls, and they had a man on. Three strikeouts, and you had retired the side. Happily, the opposing batters, apparently paralyzed by admiration, never swung at a good pitch. Unfortunately, though, you had to keep facing them; the retired side resurrected immediately from its ashes, fresh and vigorous, while you grew delirious—nutsy, that is, from fielding a bouncing ball every other second and then stilling your heart and blinking the blood from your eyes so you could concentrate on the pitch.

Amy's friend Tibby had an older brother, named Ricky; he was younger than I was, but available. We had no laughing friendship, such as I enjoyed with Pin Ford, but instead a working relationship: we played a two-handed baseball game. Tibby and Ricky's family lived secluded at the high dead end of Richland Lane. Their backyard comprised several kempt and gardened acres. It was here in the sweet mown grass, here between the fruit trees and the rhubarb patch, that we passed long, hot afternoons pitching a baseball. Ricky was a sober, good-looking boy, very dark; his father was a surgeon.

We each pitched nine innings. The other caught, hunkered down, and called each pitch a ball or a strike. That was the essence of it: Catcher called it. Four walks scored a side. Three outs retired a side, and the catcher's side came on to pitch.

This was practically the majors. You had a team to root for, a team that both received pitches and dished them out. You kept score. The pitched ball came back right to you—after a proper, rhythmical interval. You had a real squatting catcher. Best, you had a baseball.

The game required the accuracy I was always working on. It also required honor. If when you were catching you made some iffy calls, you would be sorry when it was your turn to pitch. Ricky and I were, in this primitive sense, honorable. The tag ends of summer—before or after camp, before or after Lake Erie—had thrown us together for this one activity, this chance to do some pitching. We shared a catcher's mitt every inning; we pitched at the catcher's mitt. I threw as always by imagining my whole body hurled into the target; the rest followed naturally. I had one pitch, a fast ball. I couldn't control the curve. When the game was over, we often played another. Then we thanked each other formally, drank some hot water from a garden hose, and parted—like, perhaps, boys.

On Tuesday summer evenings I rode my bike a mile down Braddock Avenue to a park where I watched Little League teams play ball. Little League teams did not accept girls, a ruling I looked into for several years in succession. I parked my bike and hung outside the chain-link fence and watched and rooted and got mad and hollered, "Idiot, catch the ball!" "Play's at first!" Maybe some coach would say, "Okay, sweetheart, if you know it all, you go in there." I thought of disguising myself. None of this was funny. I simply wanted to play the game earnestly, on a diamond, until it was over, with eighteen players who knew what they were doing, and an umpire. My parents were sympathetic, if amused, and not eager to make an issue of it.

At school we played softball. No bunting, no stealing. I had settled on second base, a spot Bill Mazeroski would later sanctify: lots of action, lots of talk, and especially a chance to turn the double play. Dumb softball: so much better than no ball at all, I reluctantly grew to love it. As I got older, and the prospect of having anything to do with young Ricky up the street became out of the question, I had to remind myself, with all loyalty and nostalgia, how a baseball, a real baseball, felt.

A baseball weighted your hand just so, and fit it. Its red stitches, its good leather and hardness like skin over bone, seemed to call forth a skill both easy and precise. On the catch—the grounder, the fly, the line drive—you could snag a baseball in your mitt, where it stayed, snap, like a mouse locked in its trap, not like some pumpkin of a softball you merely halted, with a terrible sound like a splat. You could curl your fingers around a baseball, and throw it in a straight line. When you hit it with a bat it cracked—and your heart cracked, too, at the sound. It took a grass stain nicely, stayed round, smelled good, and lived lashed in your mitt all winter, hibernating.

There was no call for overhand pitches in softball; all my training was useless. I was playing with twenty-five girls, some of whom did not, on the face of it, care overly about the game at hand. I waited out by second and hoped for a play to the plate.

A TORNADO HIT OUR NEIGHBORHOOD one morning. Our neighborhood was not only leafy Richland Lane and its hushed side streets, but also Penn Avenue, from which Richland Lane loftily arose. Old Penn Avenue was a messy, major thoroughfare still cobblestoned in the middle lanes, and full of stoplights and jammed traffic. There were drugstores there, old apartment buildings, and some old mansions. Penn Avenue was the city—tangled and muscular, a broad and snarled fist. The tornado broke all the windows in the envelope factory on Penn Avenue and ripped down mature oaks and maples on Richland Lane and its side streets—trees about which everyone would make, in my view, an unconscionable fuss, not least perhaps because they would lie across the streets for a week.

After the tornado passed I roamed around and found a broken power line. It banged violently by the Penn Avenue curb; it was shooting sparks into the street. I couldn't bring myself to leave the spot.

The power line was loosing a fireball of sparks that melted the asphalt. It was a thick twisted steel cable usually strung overhead along Penn Avenue; it carried power—4,500 kilovolts of it—from Wilkinsburg ("City of Churches") to major sections of Pittsburgh, to Homewood and Brushton, Shadyside, and Squirrel Hill.

It was melting a pit for itself in the street. The live wire's hundred twisted ends spat a thick sheaf of useless yellow sparks that hissed. The sparks were cooking the asphalt gummy; they were burning a hole. I watched the cable relax and sink into its own pit; I watched the yellow sparks pool and crackle around the cable's torn end and splash out of the pit and over the asphalt in a stream toward the curb and my shoes. My bare shins could feel the heat. I smelled tarry melted asphalt and steel so hot it smoked.

"If you touch that," my father said, needlessly, "you're a goner."

I had gone back to the house to get him so he could see this violent sight, this cable all but thrashing like a cobra and shooting a torrent of sparks.

While the tornado itself was on—while the buckeye trees in our yard were coming apart—Mother had gathered Amy and Molly and held them with her sensibly away from the windows; she urged my father and me to join them. Father had recently returned from his river trip and was ensconced tamed in the household again. And here was a pleasant, once-in-a-lifetime tornado, the funnel of which touched down, in an almost delicate point, like a bolt of lightning, on our very street. He and I raced from window to window and watched; we saw the backyard sycamore smash the back-porch roof; we saw the air roaring and blowing full of sideways-flying objects, and saw the leafy buckeye branches out front blow white and upward like skirts.

"With your taste for natural disaster," Mother said to me later, "you should try to arrange a marriage with the head of the International Red Cross."

Now the torn cable lay near the curb, away from traffic. Its loose power dissipated in the air, a random destructiveness. If you touched it, you would turn into Reddy Kilowatt. Your skin would wiggle up in waves like an electrified cat's in a cartoon; your hair would rise stiff from your head; anyone who touched you by mistake would stick to you wavy-skinned and paralyzed. You would be dead but still standing, the power surging through your body in electrical imitation of life. Passersby would have to knock you away from the current with planks.

Father placed a ring of empty Coke bottles around the hissing power line and went back home to call Duquesne

Light. I stayed transfixed. Other neighborhood children showed up, looked at the cable shooting sparks, and wandered away to see the great killed trees. I stood and watched the thick billion bolts swarm in the street. The cable was as full as a waterfall, never depleted; it dug itself a pit in which the yellow sparks spilled like water. I stayed at the busy Penn Avenue curb all day staring, until, late in the afternoon, someone somewhere turned off the juice.

Streetcars ran on Penn Avenue. Streetcars were orange, clangy, beloved things—loud, jerky, and old. They were powerless beasts compelled to travel stupidly with their wheels stuck in the tracks below them. Each streetcar had one central headlight, which looked fixedly down its tracks and nowhere else. The single light advertised to drivers at night that something was coming that couldn't move over. When a streetcar's tracks and wires rounded a corner, the witless streetcar had to follow. Its heavy orange body bulged out and blocked two lanes; any car trapped beside it had to cringe stopped against the curb until it passed.

Sometimes a car parked at the curb blocked a streetcar's route. Then the great beast sounded its mournful bell: it emitted a long-suffering, monotonous bong . . . bong . . . bong . . . and men and women on the sidewalk shook their heads sympathetically at the motorman inside, the motorman more inferred than seen through the windshield's bright reflections.

Penn Avenue smelled of gasoline, exhaust fumes, trees' sweetness in the spring, and, year round, burnt grit. On the blocks from Lang to Richland Lane were buildings in wild assortment: two drugstores, Henry Clay Frick's mansion with his old daughter somewhere inside, a dark working-class bar called the Evergreen Café, a corner grocery store, the envelope factory, a Westinghouse plant, some old apartment buildings, and a parklike Presbyterian seminary.

You walked on sidewalks whose topography was as intricate as Pittsburgh's, and as hilly. Frost-heaved peaks of cement arose, broke, and, over years, subsided again like Appalachians beside deep pits in which clean grass grew from what looked like black grease. Every long once in a while, someone repaired the sidewalk, to the tune of four or five squares' worth. The sidewalks were like greater Pittsburgh in this, too—cut into so many parts, so many legal divisions, that no one was responsible for all of it, and it all crumbled.

It was your whole body that knew those sidewalks and streets. Your bones ached with them; you tasted their hot dust in your bleeding lip; their gravel worked into your palms and knees and stayed, blue under the new skin that grew over it.

You rode your bike across Penn Avenue with the light: a lane of asphalt, a sunken streetcar track just the width of a thin bike wheel, a few feet of brown cobblestones, another streetcar track, more cobblestones or some cement, more tracks, and another strip of asphalt. The old cobblestones were pale humpy ovals like loaves. When you rode your bike over them, you vibrated all over. A particularly long humpy cobblestone could knock you down in a twinkling if it caught your bike's front wheel. So could the streetcar's tracks, and they often did; your handlebars twisted in your hands and threw you like a wrestler. So you had to pay attention, alas, and could not simply coast along over cobblestones, blissfully vibrating all over. Now the city was replacing all the cobblestones, block by block. The cobblestones had come from Pittsburgh's riverbeds. In the nineteenth century, children had earned pennies by dragging them up from the water and selling them to paving contractors. They had been a great and late improvement on mud.

The streetcars' overhead network of wires made of Penn Avenue a loose-roofed tunnel. The wires cut the sky into rectangles inside which you could compose various views as you walked. Here were a yellow brick apartment top and some flattened fair-weather clouds; here were green sycamore leaves in the foreground, and a faded orange rooftop advertising sign, and a yellow streetlight, and a slab of neutral sky.

Streetcars traveled with their lone trolley sticks pushed up by springs into these overhead wires. A trolley stick carried a trolley wheel; the trolley wheel rolled along the track of hot electric wire overhead as the four wheels rolled along the cold grooved track below. At night, and whenever it rained, the streetcars' trolleys sparked. They shot a radiant fistful of sparks at every crossing of wires. Sometimes a streetcar accidentally "threw the trolley." Bumping over a switch or rounding a bend, the trolley lost the wire and the spring-loaded stick flew up and banged its bare side crazily against the hot wire. Big yellow sparks came crackling into the sky and fell glowing toward the roofs of cars. The motorman had to brake the streetcar, go around to its rear, and haul the wayward, sparking trolley stick down with a rope. This happened so often that there was a coil of rope for that purpose at the streetcar's stern, neat and cleated like a halyard on a mast.

So the big orange streetcars clanged and spat along; they stopped and started, tethered to their wires overhead and trapped in their grooves below. Every day at a hundred intersections they locked horns with cars that blocked their paths—cars driven by insensible, semiconscious people, people who had just moved to town, teenagers learning to drive, the dread Ohio drivers, people sunk in rapturous conversation.

"Bong bong," bleated the stricken streetcar, "bong," and its passengers tried to lean around to see what was holding it up, and its berserk motorman gestured helplessly, furiously, at the dumb dreaming car—a shrug, a wave, a fist:

> I'm a streetcar!
> What can I do?
> What can I do
> but wait for you jerks
> to figure out that I'm a streetcar!

I tried to kill a streetcar by overturning it.

Pin Ford and I were hiding under a purple beech tree on the lawn of the Presbyterian seminary on Penn Avenue.

Through the beech's low dense branches she and I could

make out Penn Avenue's streetcar lanes. It was midafternoon. Now a streetcar was coming toward us. We had been waiting. We had just stuck a stone in the streetcar track. This one seemed like a stone big enough to throw it over. Would the streetcar go over? Did we hope it would go over? We spotted its jiggling trolley stick first, high above the roofs of cars. Then we saw its round orange shoulder, humped like a cobblestone, and its lone simple eye. I pressed a thumb and finger between ribs on both sides of my breastbone, to try to calm myself.

It had started with pennies. A streetcar's wheel could slick a penny and enlarge it to a stripe. What would it do to a stone? It would crunch and crumble a stone. How big a stone? We ran between moving cars and placed ever bigger stones in the streetcar track; we ran back under the beech tree to watch.

This last stone was a coarse gray conglomerate, five inches by two by two. Was it reinforced concrete? Through the low-slung beech boughs we saw the streetcar draw nigh; we covered our lower faces with our hands.

The streetcar hit the stone audibly and rose like a beached whale. Its big orange body faltered in the air, heaved toward the lane of cars beside it, trembled, and finally fell down on its track and broke the stone. And went on, bumping again only slightly when the rear wheel went over it. Pin Ford and I lay low.

In that instant while the streetcar stopped upraised over its track like an animal bewildered, while it swayed over the cars' lane and hung on its side and its trolley stick dangled askew, I saw it continue its roll; I saw precisely which cars it would fall on, and which dim people silhouetted inside the cars and the streetcar would be the most surprised. I saw, too, in that clear instant, that if the streetcar did derail, I would have to come forward and give myself up to the police, and do time, and all that, for the alternative was living all the rest of life on the lam.

What can we make of the inexpressible joy of children? It is a kind of gratitude, I think—the gratitude of the ten-year-old who wakes to her own energy and the brisk challenge of the world. You thought you knew the place and all its routines, but you see you hadn't known. Whole stacks at the library held books devoted to things you knew nothing about. The boundary of knowledge receded, as you poked about in books, like Lake Erie's rim as you climbed its cliffs. And each area of knowledge disclosed another, and another. Knowledge wasn't a body, or a tree, but instead air, or space, or being—whatever pervaded, whatever never ended and fitted into the smallest cracks and the widest space between stars.

Any way you cut it, colors and shadows flickered from multiple surfaces. Just enough work had already been done on everything—moths, say, or meteorites—to get you started and interested, but not so much there was nothing left to do. Often I wondered: was it being born just now, in this century, in this country? And I thought: no, any time could have been like this, if you had the time and weren't sick; you could, especially if you were a boy, learn and do. There was joy in concentration, and the world afforded an inexhaustible wealth of projects to concentrate on. There was joy in effort, and the world resisted effort to just the right degree, and yielded to it at last. People cut Mount Rushmore into faces; they chipped here and there for years. People slowed the spread of yellow fever; they sprayed the Isthmus of Panama puddle by puddle. Effort alone I loved. Some days I would have been happy to push a pole around a threshing floor like an ox, for the pleasure of moving the heavy stone and watching my knees rise in turn.

I was running down the Penn Avenue sidewalk, revving up for an act of faith. I was conscious and self-conscious. I knew well that people could not fly—as well as anyone knows it—but I also knew the kicker: that, as the books put it, with faith all things are possible.

Just once I wanted a task that required all the joy I had. Day after day I had noticed that if I waited long enough, my strong unexpressed joy would dwindle and dissipate inside me, over many hours, like a fire subsiding, and I would at last calm down. Just this once I wanted to let it rip. Flying rather famously required the extra energy of belief, and this, too, I had in superabundance.

There were boxy yellow thirties apartment buildings on those Penn Avenue blocks, and the Evergreen Café, and Miss Frick's house set back behind a wrought-iron fence. There were some side yards of big houses, some side yards of little houses, some streetcar stops, and a drugstore from which I had once tried to heist a five-pound box of chocolates, a Whitman sampler, confusing "sampler" with "free sample." It was past all this that I ran that late fall afternoon, up old Penn Avenue on the cracking cement sidewalks—past the drugstore and bar, past the old and new apartment buildings and the long dry lawn behind Miss Frick's fence.

I ran the sidewalk full tilt. I waved my arms ever higher and faster; blood balled in my fingertips. I knew I was foolish. I knew I was too old really to believe in this as a child would, out of ignorance; instead I was experimenting as a scientist would, testing both the thing itself and the limits of my own courage in trying it miserably self-conscious in full view of the whole world. You can't test courage cautiously, so I ran hard and waved my arms hard, happy.

Up ahead I saw a business-suited pedestrian. He was coming stiffly toward me down the walk. Who could ever forget this first test, this stranger, this thin young man appalled? I banished the temptation to straighten up and walk right. He flattened himself against a brick wall as I passed flailing—although I had left him plenty of room. He had refused to meet my exultant eye. He looked away, evidently embarrassed. How surprisingly easy it was to ignore him! What I was letting rip, in fact, was my willingness to look foolish, in his eyes and in my own. Having chosen this foolishness, I was a free being. How could the world ever stop me, how could I betray myself, if I was not afraid?

I was flying. My shoulders loosened, my stride opened, my heart banged the base of my throat. I crossed Carnegie and ran up the block waving my arms. I crossed Lexington and ran up the block waving my arms.

A linen-suited woman in her fifties did meet my exultant

eye. She looked exultant herself, seeing me from far up the block. Her face was thin and tanned. We converged. Her warm, intelligent glance said she knew what I was doing—not because she herself had been a child but because she herself took a few loose aerial turns around her apartment every night for the hell of it, and by day played along with the rest of the world and took the streetcar. So Teresa of Avila checked her unseemly joy and hung on to the altar rail to hold herself down. The woman's smiling, deep glance seemed to read my own awareness from my face, so we passed on the sidewalk—a beautifully upright woman walking in her tan linen suit, a kid running and flapping her arms—we passed on the sidewalk with a look of accomplices who share a humor just beyond irony. What's a heart for?

I crossed Homewood and ran up the block. The joy multiplied as I ran—I ran never actually quite leaving the ground—and multiplied still as I felt my stride begin to fumble and my knees begin to quiver and stall. The joy multiplied even as I slowed bumping to a walk. I was all but splitting, all but shooting sparks. Blood coursed free inside my lungs and bones, a light-shot stream like air. I couldn't feel the pavement at all.

I was too aware to do this, and had done it anyway. What could touch me now? For what were the people on Penn Avenue to me, or what was I to myself, really, but a witness to any boldness I could muster, or any cowardice if it came to that, any giving up on heaven for the sake of dignity on earth? I had not seen a great deal accomplished in the name of dignity, ever.

One sunday afternoon Mother wandered through our kitchen, where Father was making a sandwich and listening to the ball game. The Pirates were playing the New York Giants at Forbes Field. In those days, the Giants had a utility infielder named Wayne Terwilliger. Just as Mother passed through, the radio announcer cried—with undue drama—"Terwilliger bunts one!"

"Terwilliger bunts one?" Mother cried back, stopped short. She turned. "Is that English?"

"The player's name is Terwilliger," Father said. "He bunted."

"That's marvelous," Mother said. " 'Terwilliger bunts one.' No wonder you listen to baseball. 'Terwilliger bunts one.' "

For the next seven or eight years, Mother made this surprising string of syllables her own. Testing a microphone, she repeated, "Terwilliger bunts one"; testing a pen or a typewriter, she wrote it. If, as happened surprisingly often in the course of various improvised gags, she pretended to whisper something else in my ear, she actually whispered, "Terwilliger bunts one." Whenever someone used a French phrase, or a Latin one, she answered solemnly, "Terwilliger bunts one." If Mother had had, like Andrew Carnegie, the opportunity to cook up a motto for a coat of arms, hers would have read simply and tellingly, "Terwilliger bunts one." (Carnegie's was "Death to Privilege.")

She served us with other words and phrases. On a Florida trip, she repeated tremulously, "That . . . is a royal poinciana." I don't remember the tree; I remember the thrill in her voice. She pronounced it carefully, and spelled it. She also liked to say "portulaca."

The drama of the words "Tamiami Trail" stirred her, we learned on the same Florida trip. People built Tampa on one coast, and they built Miami on another. Then—the height of visionary ambition and folly—they piled a slow, tremendous road through the terrible Everglades to connect them. To build the road, men stood sunk in muck to their armpits. They fought off cottonmouth moccasins and six-foot alligators. They slept in boats, wet. They blasted muck with dynamite, cut jungle with machetes; they laid logs, dragged drilling machines, hauled dredges, heaped limestone. The road took fourteen years to build up by the shovelful, a Panama Canal in reverse, and cost hundreds of lives from tropical, mosquito-carried diseases. Then, capping it all, some genius thought of the word Tamiami: they called the road from Tampa to Miami, this very road under our spinning wheels, the Tamiami Trail. Some called it Alligator Alley. Anyone could drive over this road without a thought.

Hearing this, moved, I thought all the suffering of road building was worth it (it wasn't my suffering), now that we had this new thing to hang these new words on—Alligator Alley for those who liked things cute, and, for connoisseurs like Mother, for lovers of the human drama in all its boldness and terror, the Tamiami Trail.

Back home, Mother cut clips from reels of talk, as it were, and played them back at leisure. She noticed that many Pittsburghers confuse "leave" and "let." One kind relative brightened our morning by mentioning why she'd brought her son to visit: "He wanted to come with me, so I left him." Mother filled in Amy and me on locutions we missed. "I can't do it on Friday," her pretty sister told a crowded dinner party, "because Friday's the day I lay in the stores."

(All unconsciously, though, we ourselves used some pure Pittsburghisms. We said "tele pole," pronounced "telly pole," for that splintery sidewalk post I loved to climb. We said "slippy"—the sidewalks are "slippy." We said, "That's all the farther I could go." And we said, as Pittsburghers do say, "This glass needs washed," or "The dog needs walked"—a usage our father eschewed; he knew it was not standard English, nor even comprehensible English, but he never let on.)

"Spell 'poinsettia,' " Mother would throw out at me, smiling with pleasure. "Spell 'sherbet.' " The idea was not to make us whizzes, but, quite the contrary, to remind us—and I, especially, needed reminding—that we didn't know it all just yet.

"There's a deer standing in the front hall," she told me one quiet evening in the country.

"Really?"

"No. I just wanted to tell you something once without your saying, 'I know.' "

Supermarkets in the middle 1950s began luring, or bothering, customers by giving out Top Value Stamps or Green Stamps. When, shopping with Mother, we got to the head of the checkout line, the checker, always a young man, asked, "Save stamps?"

"No," Mother replied genially, week after week, "I build model airplanes." I believe she originated this line. It took me years to determine where the joke lay.

Anyone who met her verbal challenges she adored. She had surgery on one of her eyes. On the operating table, just before she conked out, she appealed feelingly to the surgeon, saying, as she had been planning to say for weeks, "Will I be able to play the piano?" "Not on me," the surgeon said. "You won't pull that old one on me."

It was, indeed, an old one. The surgeon was supposed to answer, "Yes, my dear, brave woman, you will be able to play the piano after this operation," to which Mother intended to reply, "Oh, good, I've always wanted to play the piano." This pat scenario bored her; she loved having it interrupted. It must have galled her that usually her acquaintances were so predictably unalert; it must have galled her that, for the length of her life, she could surprise everyone so continually, so easily, when she had been the same all along. At any rate, she loved anyone who, as she put it, saw it coming, and called her on it.

She regarded the instructions on bureaucratic forms as straight lines. "Do you advocate the overthrow of the United States government by force or violence?" After some thought she wrote, "Force." She regarded children, even babies, as straight men. When Molly learned to crawl, Mother delighted in buying her gowns with drawstrings at the bottom, like Swee'pea's, because, as she explained energetically, you could easily step on the drawstring without the baby's noticing, so that she crawled and crawled and crawled and never got anywhere except into a small ball at the gown's top.

When we children were young, she mothered us tenderly and dependably; as we got older, she resumed her career of anarchism. She collared us into her gags. If she answered the phone on a wrong number, she told the caller, "Just a minute," and dragged the receiver to Amy or me, saying, "Here, take this, your name is Cecile," or, worse, just, "It's for you." You had to think on your feet. But did you want to perform well as Cecile, or did you want to take pity on the wretched caller?

During a family trip to the Highland Park Zoo, Mother and I were alone for a minute. She approached a young couple holding hands on a bench by the seals, and addressed the young man in dripping tones: "Where have you been? Still got those baby-blue eyes; always did slay me. And this"—a swift nod at the dumbstruck young woman, who had removed her hand from the man's—"must be the one you were telling me about. She's not so bad, really, as you used to make out. But listen, you know how I miss you, you know where to reach me, same old place. And there's Ann over there—see how she's grown? See the blue eyes?"

And off she sashayed, taking me firmly by the hand, and leading us around briskly past the monkey house and away. She cocked an ear back, and both of us heard the desperate man begin, in a high-pitched wail, "I swear, I never saw her before in my life. . . ."

On a long, sloping beach by the ocean, she lay stretched out sunning with Father and friends, until the conversation gradually grew tedious, when without forethought she gave a little push with her heel and rolled away. People were stunned. She rolled deadpan and apparently effortlessly, arms and legs extended and tidy, down the beach to the distant water's edge, where she lay at ease just as she had been, but half in the surf, and well out of earshot.

She dearly loved to fluster people by throwing out a game's rules at whim—when she was getting bored, losing in a dull sort of way, and when everybody else was taking it too seriously. If you turned your back, she moved the checkers around on the board. When you got them all straightened out, she denied she'd touched them; the next time you turned your back, she lined them up on the rug or hid them under your chair. In a betting rummy game called Michigan, she routinely played out of turn, or called out a card she didn't hold, or counted backward, simply to amuse herself by causing an uproar and watching the rest of us do double takes and have fits. (Much later, when serious suitors came to call, Mother subjected them to this fast card game as a trial by ordeal; she used it as an intelligence test and a measure of spirit. If the poor man could stay a round without breaking down or running out, he got to marry one of us, if he still wanted to.)

She excelled at bridge, playing fast and boldly, but when the stakes were low and the hands dull, she bid slams for the devilment of it, or raised her opponents' suit to bug them, or showed her hand, or tossed her cards in a handful behind her back in a characteristic swift motion accompanied by a vibrantly innocent look. It drove our stolid father crazy. The hand was over before it began, and the guests were appalled. How do you score it, who deals now, what do you do with a crazy person who is having so much fun? Or they were down seven, and the guests were appalled. "Pam!" "Dammit,

Pam!" He groaned. What ails such people? What on earth possesses them? He rubbed his face.

She was an unstoppable force; she never let go. When we moved across town, she persuaded the U.S. Post Office to let her keep her old address—forever—because she'd had stationery printed. I don't know how she did it. Every new post office worker, over decades, needed to learn that although the Doaks' mail is addressed to here, it is delivered to there.

Mother's energy and intelligence suited her for a greater role in a larger arena—mayor of New York, say—than the one she had. She followed American politics closely; she had been known to vote for Democrats. She saw how things should be run, but she had nothing to run but our household. Even there, small minds bugged her; she was smarter than the people who designed the things she had to use all day for the length of her life.

"Look," she said. "Whoever designed this corkscrew never used one. Why would anyone sell it without trying it out?" So she invented a better one. She showed me a drawing of it. The spirit of American enterprise never faded in Mother. If capitalizing and tooling up had been as interesting as theorizing and thinking up, she would have fired up a new factory every week, and chaired several hundred corporations.

"It grieves me," she would say, "it grieves my heart," that the company that made one superior product packaged it poorly, or took the wrong tack in its advertising. She knew, as she held the thing mournfully in her two hands, that she'd never find another. She was right. We children wholly sympathized, and so did Father; what could she do, what could anyone do, about it? She was Samson in chains. She paced.

She didn't like the taste of stamps so she didn't lick stamps; she licked the corner of the envelope instead. She glued sandpaper to the sides of kitchen drawers, and under kitchen cabinets, so she always had a handy place to strike a match. She designed, and hounded workmen to build against all norms, doubly wide kitchen counters and elevated bathroom sinks. To splint a finger, she stuck it in a lightweight cigar tube. Conversely, to protect a pack of cigarettes, she carried it in a Band-Aid box. She drew plans for an over-the-finger toothbrush for babies, an oven rack that slid up and down, and—the family favorite—Lendalarm. Lendalarm was a beeper you attached to books (or tools) you loaned friends. After ten days, the beeper sounded. Only the rightful owner could silence it.

She repeatedly reminded us of P. T. Barnum's dictum: You could sell anything to anybody if you marketed it right. The adman who thought of making Americans believe they needed underarm deodorant was a visionary. So, too, was the hero who made a success of a new product, Ivory soap. The executives were horrified, Mother told me, that a cake of this stuff floated. Soap wasn't supposed to float. Anyone would be able to tell it was mostly whipped-up air. Then some inspired adman made a leap: Advertise that it floats. Flaunt it. The rest is history.

She respected the rare few who broke through to new ways. "Look," she'd say, "here's an intelligent apron." She called upon us to admire intelligent control knobs and intelligent pan handles, intelligent andirons and picture frames and knife sharpeners. She questioned everything, every pair of scissors, every knitting needle, gardening glove, tape dispenser. Hers was a restless mental vigor that just about ignited the dumb household objects with its force.

Torpid conformity was a kind of sin; it was stupidity itself, the mighty stream against which Mother would never cease to struggle. If you held no minority opinions, or if you failed to risk total ostracism for them daily, the world would be a better place without you.

Always I heard Mother's emotional voice asking Amy and me the same few questions: Is that your own idea? Or somebody else's? "*Giant* is a good movie," I pronounced to the family at dinner. "Oh, really?" Mother warmed to these occasions. She all but rolled up her sleeves. She knew I hadn't seen it. "Is that your considered opinion?"

She herself held many unpopular, even fantastic, positions. She was scathingly sarcastic about the McCarthy hearings while they took place, right on our living-room television; she frantically opposed Father's wait-and-see calm. "We don't know enough about it," he said. "I do," she said. "I know all I need to know."

She asserted, against all opposition, that people who lived in trailer parks were not bad but simply poor, and had as much right to settle on beautiful land, such as rural Ligonier, Pennsylvania, as did the oldest of families in the finest of hidden houses. Therefore, the people who owned trailer parks, and sought zoning changes to permit trailer parks, needed our help. Her profound belief that the country-club pool sweeper was a person, and that the department-store saleslady, the bus driver, telephone operator, and house-painter were people, and even in groups the steelworkers who carried pickets and the Christmas shoppers who clogged intersections were people—this was a conviction common enough in democratic Pittsburgh, but not altogether common among our friends' parents, or even, perhaps, among our parents' friends.

Opposition emboldened Mother, and she would take on anybody on any issue—the chairman of the board, at a cocktail party, on the current strike; she would fly at him in a flurry of passion, as a songbird selflessly attacks a big hawk.

"Eisenhower's going to win," I announced after school. She lowered her magazine and looked me in the eyes: "How do you know?" I was doomed. It was fatal to say, "Everyone says so." We all knew well what happened. "Do you consult this Everyone before you make your decisions? What if Everyone decided to round up all the Jews?" Mother knew there was no danger of cowing me. She simply tried to keep us all awake. And in fact it was always clear to Amy and me, and to Molly when she grew old enough to listen, that if our classmates came to cruelty, just as much as if the neighborhood or the nation came to madness, we were expected to take, and would be each separately capable of taking, a stand.

THE FRENCH AND INDIAN WAR was a war of which I, for one, reading stretched out in the bedroom, couldn't get enough. The names of the places were a litany: Fort Ticonderoga on the Hudson, Fort Vincennes on the Wabash. The names of the people were a litany: the Sieur de Contrecoeur; the Marquis de Montcalm; Major Robert Rogers of the Rangers; the Seneca Chief Half-King.

How witless in comparison were the clumsy wars of Europe: on this open field at nine o'clock sharp, soldiers in heavy armor, dragged from their turnip patches in feudal obedience to Lord So-and-So, met in long ranks the heavily armored men owned or paid for by Lord So-and-So, and defeated them by knocking them over like ninepins. What was at stake? A son's ambition, or an earl's pride.

In the French and Indian War, and the Indian wars, a whole continent was at stake, and it was hard to know who to root for as I read. The Indians were the sentimental favorites, but they were visibly cruel. The French excelled at Indian skills and had the endearing habit of singing in boats. But if they won, we would all speak French, which seemed affected in the woods. The Scotch-Irish settlers and the English army were very uneasy allies, but their cruelties were invisible to me, and their partisans wrote all the books that fell into my hands.

It all seemed to take place right here, here among the blossoming rhododendrons outside the sunporch windows just below our bedroom, here in the Pittsburgh forest that rose again from every vacant lot, every corner of every yard the mower missed, every dusty crack in the sidewalk, every clogged gutter on the roof—an oak tree, a sycamore, a mountain ash, a pine.

For here, on the tip of the point where the three rivers met, the French built Fort Duquesne. It linked French holdings on the Great Lakes to their settlement at New Orleans. It was 1754; the forest was a wilderness. From Fort Duquesne the French set their Indian allies to raiding far-flung English-speaking settlements and homesteads. The Indians burned the farms and tortured many farm families. From Fort Duquesne the French marched out and defeated George Washington at nearby Fort Necessity. From Fort Duquesne the French marched out and defeated General Edward Braddock: Indian warriors shot from cover, which offended those British soldiers who had time to notice before they died. It was here in 1758 that General John Forbes established British hegemony over the Mississippi watershed, by driving the French from the point and building Fort Pitt.

Here our own doughty provincials in green hunting shirts fought beside regiments of rangers in buckskins, actual Highlanders in kilts, pro-English Iroquois in warpaint, and British regulars in red jackets. They came marching vividly through the virgin Pittsburgh forest; they trundled up and down the nearby mountain ridges by day and slept at night on their weapons under trees. Pioneer scouts ran ahead of them and behind them; messengers snuck into their few palisaded forts, where periwigged English officers sat and rubbed their foreheads while naked Indians in the treetops outside were setting arrows on fire to burn down the roof.

Best, it was all imaginary. That the French and Indian War took place in this neck of the woods merely enhanced its storied quality, as if that fact had been a particularly pleasing literary touch. This war was part of my own private consciousness, the dreamlike interior murmur of books.

Costumed enormous people, transparent, vivid, and bold as decals, as tall and rippling as people in dreams, shot at each other up and down the primeval woods, race against race. Just as people in myths travel rigidly up to the sky, or are placed there by some great god's fingers, to hold still forever in the midst of their loving or battles as fixed constellations of stars, so the fighting cast of the French and Indian War moved in a colorful body—locked into position in the landscape but still loading muskets or cowering behind the log door or landing canoes on a muddy shore—into books. They were fabulous and morally neutral, like everything in history, like everything in books. They were imagination's playthings: toy soldiers, toy settlers, toy Indians. They were a part of the interior life; they were private; they were my own.

In books these wars played themselves out ceaselessly; the red-warpainted Indian tomahawked the settler woman in calico, and the rangy settler in buckskin spied out the Frenchman in military braid. Whenever I opened the book, the war struck up again, like a record whose music sounded when the needle hit. The skirling of Highlanders' bagpipes came playing again, high and thin over the dry oak ridges. The towheaded pioneer schoolchildren were just blabbing their memorized psalms when from right outside the greased parchment window sounded the wild and fatal whoops of Indian warriors on a raid.

The wild and fatal whoops, the war whoops of the warriors, the red warriors whooping on a raid. It was a delirium. The tongue diddled the brain. Private life, book life, took place where words met imagination without passing through world.

I could dream it all whenever I wanted—and how often I wanted to dream it! Fiercely addicted, I dosed myself again and again with the drug of the dream.

Parents have no idea what the children are up to in their bedrooms: They are reading the same paragraphs over and over in a stupor of violent bloodshed. Their legs are limp with horror. They are reading the same paragraphs over and over, dizzy with gratification as the young lovers find each other in the French fort, as the boy avenges his father, as the sound of muskets in the woods signals the end of the siege. They could not move if the house caught fire. They hate the actual world. The actual world is a kind of tedious plane where dwells, and goes to school, the body, the boring body which houses the eyes to read the books and houses the heart the books enflame. The very boring body seems to require an inordinately big, very boring world to keep it up, a world where you have to spend far too much time, have to *do* time like a prisoner, always looking for a chance to slip away, to escape back home to books, or escape back home to any concentration—fanciful, mental, or physical—where you can lose your self at last. Although I was hungry all the time, I could not bear to hold still and eat; it was too dull a thing to do, and had no appeal either to courage or to imagination. The blinding sway of their inner lives makes children immoral. They find things good insofar as they are thrilling, insofar as they render them ever more feverish and breathless, ever more limp and senseless on the bed.

Throughout these long, wonderful wars, I saw Indian braves behind every tree and parked car. They slunk around, fairly bursting with woodcraft. They led soldiers on miraculous escapes through deep woods and across lakes at night; they paddled their clever canoes noiselessly; they swam underwater without leaving bubbles; they called to each other like owls. They nocked their arrows silently on the brow of the hill and snuck up in their soft moccasins to the camp where the enemy lay sleeping under heavy guard. They shrieked, drew their osage bows, and never missed—all the while communing deeply with birds and deer.

I had been born too late. I would have made a dandy scout, although I was hungry all the time, because I had taught myself, with my friend Pin, to walk in the woods silently: without snapping a twig, which was easy, or stepping on a loud leaf, which was hard. Experience taught me a special, rolling walk for skulking in silence: you step down with your weight on the ball of your foot, and ease it to your heel.

The Indians who captured me would not torture me, but would exclaim at my many abilities, and teach me more, all the while feeding me handsomely. Soon I would talk to animals, become invisible, ride a horse naked and shrieking, shoot things.

I practiced traveling through the woods in Frick Park without leaving footprints. I practiced tracking people and animals, such as the infamous pedigreed dachshunds, by following sign. I knew the mark of Walter Milligan's blunt heel and the mark of Amy's sharp one. I practiced sneaking up on Mother as she repotted a philodendron, Father as he washed the car, saying, as I hoped but doubted the Indians said, "Boo."

AT SCHOOL we memorized a poem:

> Where we live and work today
> Indian children used to play—
> All about our native land
> Where the shops and houses stand.

Richland Lane was untrafficked, hushed, planted in great shade trees, and peopled by wonderfully collected children. They were sober, sane, quiet kids, whose older brothers and sisters were away at boarding school or college. Every warm night we played organized games—games that were the sweetest part of those sweet years, that long suspended interval between terror and anger.

On the quiet dead-end side street, among the still brick houses under their old ash trees and oaks, we paced out the ritual evenings. I saw us as if from above, even then, even as I stood in place living out my childhood and knowing it, aware of myself as if from above and behind, skinny and exultant on the street. We are silent, waiting or running, spread out on the pale street like chessmen, stilled as priests, relaxed and knowing. Someone hits the ball, someone silent far up the street catches it on the bounce; we move aside, clearing a path. Carefully the batter lays down the bat perpendicular to the street. Carefully the hushed player up the street rolls the ball down to the bat. The rolled ball hits the bat and flies up unpredictably; the batter misses his catch; he and the fielder switch positions. Indian Ball.

And there were no roads at all.
And the trees were very tall.

Capture the Flag was, essentially, the French and Indian War. The dead-end street (Europe) saw open combat at its fixed border. Brute strength could win. We disdained the street, although of course we had to guard its border. We fought the real war in the backyards (America)—a limitless wilderness of trees, garbage cans, thickets, back porches, and gardens, where no one knew where the two sides' territories ended, and where strategy required bold and original planning, private initiative, sneaky scouting, and courage.

If someone cheated at any game, or incurred the group's wrath in any way, the rest of us gave him, or her, Indian burns: we wrung a bare arm with both hands close together till the skin chafed. Worse—reserved for practically capital crimes—was the dreaded but admired typewriter torture, which we understood to be, in modern guise, an old Indian persuader. One of us straddled the offender, bared his or her breastbone, and lightly tapped fingertips there—very lightly, just where the skin covers the bone most closely. This light tapping does not hurt at all for the first five minutes or so.

We were nice kids who rarely resorted to torture. We played Red Rover, a variation on Prisoners' Base called Beckons Wanted, and Crack the Whip. Everything else, and parts of these games, too, smacked of Indians. By day, Pin Ford and I played at being Indians straight out. Her parents were also young, and she was my age, an only child; they lived two doors up. Pin's real name was Barbara. She was tan and blond, sturdy, smooth of skin; she was agreeable and quick to laugh. Her courage and her flair for the visual arts hadn't yet formed. She was content now to stalk the neighborhood and knock over the odd streetcar.

As Indians, Pin and I explored the wooded grounds of the Presbyterian seminary at our backyards. We made bows and arrows: we peeled and straightened deadfall sticks for arrows, and cut, stealthily, green boughs to bend for bows. With string we rigged our mothers' Chesterfield cigarette cartons over our shoulders as quivers. We shot our bows. We threw knives at targets, and played knife-throwing games. We walked as the Indians had walked, stirring no leaves, snapping no twigs. We built an Indian village, Navajo style, under the seminary's low copper beech: we baked clay bricks on slate roofing tiles set on adobe walls around a twiggy fire.

We named the trees. We searched the sky for omens, and inspected the ground for sign.

We came home and found our mothers together in our side yard by the rose garden, tanning on chaises longues. They were both thin and blond. They held silvered cardboard reflectors up to their flung-back chins. Over their closed eyelids they had placed blue eye-shaped plastic cups, joined over the nose.

THE ATTIC BEDROOM where I drew my baseball mitt was a crow's nest, a treehouse, a studio, an office, a forensic laboratory, and a fort. It interested me especially for a totemic brown water stain on a sloping plaster wall. The stain looked like a square-rigged ship heeled over in a storm. I examined this ship for many months. It was a painting, not a drawing; it had no lines, only forms awash, which rose faintly from the plaster and deepened slowly and dramatically as I watched and the seas climbed and the wind rose before anyone could furl the sails. Those distant dashes over the water—were they men sliding overboard? Were they storm petrels flying? I knew a song whose chorus asked, What did the deep sea say?

My detective work centered around the attic, and sometimes included Pin Ford. We filed information on criminal suspects in a shoe box. We got the information by hanging around the Evergreen Café on Penn Avenue and noting suspicious activity.

One dark, rainy afternoon when I was alone, I saw a case of beer inside the trunk of a man's car. If that wasn't suspicious, I didn't know what was. I was lurking just outside the drugstore, where I could see the Evergreen Café clientele without being seen. I memorized the car license number, of course, as anyone would—but my real virtue as a detective was that I could memorize the whole man, inch by inch, by means of sentences, and later reproduce the man in a drawing.

When I came home from the dark rain that afternoon I walked through floor after floor of the lighted house, wetting the golden rugs and muttering, until I got to the attic stairs and the attic itself. There I repaired to a card table under the square-rigged ship. I wrote down the suspect's car's make and license number. I wrote down my stabs at his height and age, and a description of his clothes. Then I turned on the radio, opened a cheap drawing tablet, and relaxed to the business of drawing the man who had stepped out of the Evergreen Café and revealed a case of beer in the trunk of his car.

By accident I drew a sloppy oval that looked like his head. I copied a page of these. Paying attention, I marked off some rough ratios: the crucial intervals between eye sockets, headtop, and chin. Unconsciously again, I let my hand scribble lines for features. I sat up to play back in my head certain memorized sentences: he has a wide mouth; his mouth corners fall directly beneath eyes' outer corners; forehead is round; ears are high, triangular. My dumb hand molded the recurved facial masses and shaded the eye sockets for its own pleasure with slanting parallel lines. I sat enchanted and unwitting in a trance.

What will the weather be?
Tell us, Mister Weather Man.

The radio woman enunciated her slow, terrible song. She sounded her notes delicately, as did the idiot xylophone that preceded her. A wind was rising outside. Across the attic room, the blackened windows rattled. I saw their glossed reflections on the pale walls wag. The rain battered the roof over my head, over the waterlogged ship. I heard the bare buckeye boughs hitting the house.

I was drawing the head. I shut my eyes. I could not see the man's face eidetically. That is, I could not reproduce it interiorly, study it, and discover new things, as some few people can look at a page, print it, as it were, in their memories, and read it off later. I could produce stable images only rarely. But like anyone, I could recall and almost see fleet torn fragments of a scene: a raincoat sleeve's wrinkling, a blond head bending, red-lighted rain falling on asphalt, a pesteringly interesting pattern in a cordovan shoe, which rises and floats across that face I want to see. I perceived these sights as scraps that floated like blowing tissue across some hollow interior space, some space at the arching roof of the rib cage, perhaps. I swerved to study them before they slid away.

I hoped that the sentences would nail the blowing scraps down. I hoped that the sentences would store scenes like rolls of film, rolls of film I could simply reel off and watch. But of course, the sentences did not work that way. The sentences suggested scenes to the imagination, which were no sooner repeated than envisioned, and envisioned just as poorly and just as vividly as actual memories. Here was Raggedy Ann, say, an actual memory, with her red-and-white-striped stockings and blunt black feet. And here, say, was a barefoot boy asleep in a car, his cheeks covered thinly with blood. Which was real? The barefoot boy was just as vivid. It was easier to remember a sentence than a sight, and the sentences suggested sights new or skewed. These were dim regions, these submerged caves where waters mingled. On my cheap tablet I was drawing round lips, suns, fish in schools.

Soon someone would call me for dinner. But I would not come, I suddenly realized, and I would not answer the call—ever—for I would have died of starvation. They would find me, having slid off my chair, half under the card table, lying dead on the floor. And so young.

In the blue shoe box on the card table they would find my priceless files. I had written all my data about today's suspect, drawn his face several times from several angles, and filed it all under his car's license number. When the police needed it, it was ready.

Privately I thought the reference librarian at the Homewood Library was soft in the head. The week before, she had handed me, in broad daylight, the book that contained the key to Morse code. Without a word, she watched me copy it, pocket the paper, and leave.

I knew how to keep a code secret, if she didn't. At home I memorized Morse code promptly, and burned the paper.

I had read the library's collection of popular forensic

medicine, its many books about Scotland Yard and the FBI, a dull biography of J. Edgar Hoover, and its Sherlock Holmes. I knew I was not alone in knowing Morse code. The FBI knew it, Scotland Yard knew it, and every sparks in the navy knew it. I read everything I could get about ham radios. All I needed was a receiver. I could listen in on troop maneuvers, intelligence reports, and disasters at sea. And I could rescue other hams from calamity, to which, as a class, they seemed remarkably prone.

I knew that police artists made composite drawings of criminal suspects. Witnesses to crimes selected, from a varied assortment, a stripe of crown hair, a stripe or two of forehead, a stripe of eyes, and so forth. Police artists—of whose ranks I was an oblate—made a drawing that combined these elements; newspapers published the drawings; someone recognized the suspect and called the police.

When Pin Ford and I were running low on suspects, and had run out of things to communicate in Morse code, I sat at my attic table beside the shoe box file and drew a variety of such stripes. I amused myself by combining them into new faces. So God must sit in heaven, at a card table, fingering a heap of stripes—hairlines, jawlines, brows—and joining them at whim to people a world. I began wondering if the stock of individual faces on earth through all of time is infinite.

My sweetest ambition was to see a drawing of mine on a newspaper's front page: HAS ANYBODY SEEN THIS MAN? I didn't care about reducing crime, any more than Sherlock Holmes did. I rather wished there were more crime, and closer by. What interested me was the schematic likeness, how recognizable it was, and how startlingly few things you needed to strike a resemblance. You needed only a few major proportions in the head. The soft tissues scarcely mattered; they were merely decorations that children drew. What mattered was the framing of the skull.

And so in that faraway attic, among the boughs of buckeye trees, year after year, I drew. I drew formal, sustained studies of my left hand still on the card table, of my baseball mitt, a saddle shoe. I drew from memory the faces of the people I knew, my own family just downstairs in the great house—oh, but I hated these clumsy drawings, these beloved faces so rigid on the page and lacking in tenderness and irony. (Who could analyze a numb skull when all you cared about was a lively caught glance, the pleased rising of Mother's cheek, the soft amused setting of Amy's lip, Father's imagining eye in its socket?) And I drew from memory the faces of people I saw in the streets. I formed sentences about them as I looked at them, and repeated the sentences to myself as I wandered on.

I wanted to notice everything, as Holmes had, and remember it all, as no one had before. Noticing and remembering were the route to Scotland Yard, where I intended to find my niche. They were also, more urgently, the route to the corner yard on Edgerton Avenue, to life in the house we had left and lost.

Hadn't I already forgotten the floor plan of that house where I had lived for seven years? I could see a terrifying oblong of light bend across a room's corner; I could see my mother talking on the phone in the dark stairwell, and Jo Ann Sheehy skating at night on the iced street, and the broom-closet door opening to reveal—the broom. But who could stitch these ripped remnants together? I could no longer conjure up the face of Walter Milligan, the red-haired Irish boy I had chased up and down a football field—could no longer remember his face because I had neglected to memorize it.

Noticing and remembering everything would trap bright scenes to light and fill the blank and darkening past which was already piling up behind me. The growing size of that blank and ever-darkening past frightened me; it loomed beside me like a hole in the air and battened on scraps of my life I failed to claim. If one day I forgot to notice my life, and be damned grateful for it, the blank cave would suck me up entire.

From now on, I would beat the days into my brain. Every year, every month, I vowed this vow in a different form.

But the new scenes I tried to memorize with the aid of sentences were as elusive and random as the scenes I remembered without effort. They were just as broken, trivial, capsizing, submerged. Instead of a suspect's face I saw red-lighted rain in front of a car's taillight. Instead of the schoolyard recess scene I loved, the dodgeball game I tried to memorize at one moment, and then at another—my friends and I excited and whooping—I saw a coarse cement corner, and the cyclone fence above it, and only a flash of dark green school uniforms below. Instead of my sister Molly just starting to walk I saw the smocking on her blue dress, and her stained palm. These were torn and out-of-focus scenes playing on windblown scraps. They dissolved when I tried to inspect them, or dimmed, or slid dizzyingly away, like a ship's stern yawing down the dark lee slope of a wave.

BUT HE SAID UNTO JESUS, And who is my neighbor?

And Jesus answering said, A certain man went down from Jerusalem to Jericho, and fell among thieves.

And he said unto him, Who is my neighbor?

But a certain Samaritan came where he was.

And went to him, and bound up his wounds, and brought him to an inn, and took care of him.

And he said unto him, Which now, thinkest thou, was neighbor unto him that fell among the thieves?

And he said unto him, Who IS my neighbor?

And Jesus answering said, A certain man went down from Jerusalem to Jericho, and fell among thieves.

Who IS my neighbor?

Then said Jesus unto him, Go, and do thou likewise.

This was my "Terwilliger bunts one." This and similar fragments of Biblical language played in my head like a record on which the needle has stuck, played at the back of my mind and moved at the root of my tongue and sounded deep in my ears without surcease. Who IS my neighbor?

Every July for four years, Amy and I trotted off to a Presbyterian church camp. It was cheap, wholesome, and nearby. There we were happy, loose with other children in cabins under pines. If our parents had known how pious and low church this camp was, they would have yanked us. We memorized Bible chapters, sang rollicking hymns around the clock, held nightly devotions including extemporaneous prayers, and filed out of the woods to chapel twice on Sundays dressed in white shorts. The faith-filled theology there was only half a step out of a tent; you could still smell the sawdust.

We met all sorts of girls at camp. There were a dozen girls from an orphanage, who had never been adopted. Among these I admired an older girl named Liz—a large-framed, bony girl with dry blond curls and high red cheekbones, who wore a wool lumberjack shirt. Every Sunday night, gathered in our bare old rec hall of a chapel, we children could request a favorite hymn if we could recite a Bible verse. Year after year, big Liz returned unadopted to camp and, Sunday after Sunday, requested "No One Ever Cared for Me Like Jesus."

I had a head for religious ideas. They were the first ideas I ever encountered. They made other ideas seem mean.

For what shall it profit a man, if he shall gain the whole world and lose his own soul? And lose his own soul? And lose his own soul? Know ye that the Lord he is God: it is he that hath made us, and not we ourselves; we are his people, and the sheep of his pasture. Arise, and take up thy bed, and walk. And he said unto him, WHO IS MY NEIGHBOR?

Who shall ascend into the hill of the Lord? or who shall stand in his holy place? He that hath clean hands, and a pure heart; who hath not lifted up his soul unto vanity, nor sworn deceitfully.

The earth is the Lord's, and the fulness thereof; the world, and they that dwell therein. The heavens declare the glory of God; and the firmament sheweth his handywork. Verily I say unto you, that one of you shall betray me.

Every summer we memorized these things at camp. Every Sunday in Pittsburgh we heard these things in Sunday school. Every Thursday we studied these things, and memorized them, too (strictly as literature, they said), at school. I had miles of Bible in memory: some perforce, but most by hap, like the words to songs. There was no corner of my brain where you couldn't find, among the files of clothing labels and heaps of rocks, among the swarms of protozoans and shelves of novels, whole tapes and snarls and reels of Bible. Later, before I left Pittsburgh for college, I would write several poems in deliberate imitation of its sounds, those repeated feminine endings followed by thumps, or those long hard beats followed by softness. Selah.

The Bible's was an unlikely, movie-set world alongside our world. Light-shot and translucent in the pallid Sunday-school watercolors on the walls, stormy and opaque in the dense and staggering texts they read us placidly, sweet-mouthed and earnest, week after week, this world interleaved our waking world like dream.

The adult members of society adverted to the Bible unreasonably often. What arcana! Why did they spread this scandalous document before our eyes? If they had read it, I thought, they would have hid it. They didn't recognize the vivid danger that we would, through repeated exposure, catch a case of its wild opposition to their world. Instead they bade us study great chunks of it, and think about those chunks, and commit them to memory, and ignore them. By dipping us children in the Bible so often, they hoped, I think, to give our lives a serious tint, and to provide us with quaintly magnificent snatches of prayer to produce as charms while, say, being mugged for our cash or jewels.

In Sunday school at the Shadyside Presbyterian Church, the handsome father of rascal Jack from dancing school, himself a vice-president of Jones & Laughlin, whose wife was famous at the country club for her tan, held a birch pointer in his long fingers and shyly tapped the hanging paper map, shyly because he could see we weren't listening. Who would listen to this? Why on earth were we here? There in blue and yellow and green were Galilee, Samaria itself, and Judaea, he said—and I pretended to pay attention as a courtesy—the Sea of Galilee, the river Jordan, and the Dead Sea. I saw on the hanging map the coasts of Judaea by the far side of Jordan, on whose unimaginable shores the pastel Christ had maybe uttered such cruel, stiff, thrilling words: Sell whatsoever thou hast.

James and John, the sons of Zebedee, he made them fishers of men. And he came to the Lake of Gennesaret, and he came to Capernaum. And he withdrew in a boat. And a certain man went down from Jerusalem to Jericho. See it here on the map? Down. He went down, and fell among thieves.

And the swine jumped over the cliff.

And the voice cried, Samuel, Samuel. And the wakened boy Samuel answered, Here am I. And at last he said, Speak.

Hear O Israel, the Lord is one.

And Peter said, I know him not; I know him not; I know him not. And the rich young ruler said, What must I do? And the woman wiped his feet with her hair. And he said, Who touched me?

And he said, Verily, verily, verily, verily; life is not a dream. Let this cup pass from me. If it be thy will, of course, only if it be thy will.

I GOT MY ROCK COLLECTION from our grandparents' paper boy. He handed it to me in three heavy grocery bags; he said he had no time for a rock collection. Amy and I visited Oma and Company every Friday; while Mary cooked dinner, I roamed their solemn neighborhood, where our family would, as it happened, soon live ourselves. The indigenous children kept mum inside their stone houses; the paper boy—having pedaled his thick black bike, plus rocks, up from an Italian neighborhood down the hill—was the only sign of life.

The paper boy got the rock collection from a solitary old man named Downey, who until recently had lived just up the street from my grandparents. Mr. Downey had collected the rocks from all over. He had given them to the paper boy, in the grocery bags, explaining that he knew no one else. Then he had died. The paper boy, who was kind but very busy, did not remember the names of any of the rocks except the stalactites; he recalled, not helpfully, that Mr. Downey had found them in a cave. The stalactites were sorry-looking at their broken ends: sharp, yellow, and hollow, like fallen deciduous teeth.

Now I had these rocks. They were yellow, green, blue, and red. Most were the size of half-bricks. One small white wafer had blue stone stars. Some were knobby, some grainy, some slick. There was a shining brown mineral the color of shoe polish; its cubed crystals made a scratchy chunk. There was a rusty cluster of petrified roses. There was a frozen froth of platinum bubbles. It was a safe bet all these rocks had names.

From the Homewood Library's children's books I could learn only the vaguest, overamazed stories of "the earth's crust," which didn't interest me. What were all these yellow and blue rocks in my room, and why did I never find any rocks so various and sharp? From library adult books I got the true dope, and it was a long story, which involved me in a project less like bird-watching or stamp collecting than like life in a forensic laboratory. The books taught me to identify the rocks. They also lent me a vision of things, and informed me about a bizarre set of people.

You got Frederick H. Pough's *Field Guide to Rocks and Minerals*. Using this and other books, you identified your rocks one by one, keying them out as you key out plants with Gray's *Manual*, through a series of diagnostic tests.

You determined, for instance, where your rock fit on Mohs' scale of hardness. Mr. Mohs (Herr Mohs, actually) had devised a series of homespun tests for rock hardness, much as Mr. Beaufort had dreamed up homespun tests for wind force. Does smoke rise straight from the chimney? The wind is not blowing. Is your house falling over? It's blowing force ten. Number one on Mohs' scale was soft rock, to wit, talc. Can you crumble it in your fingers? It's soft. What you have there is talcum powder. Can it scratch a fingernail, a copper penny, a pane of glass, and a knife blade? It's quartz. You can scratch quartz with topaz, ruby, and diamond. If it makes your diamond saw clog, it's a meteorite.

You subjected your rocks to scratch tests. You procured a piece of bathroom tile (always, in the books, hexagonal, such as you find in old New York bathrooms and nowhere else) and stroked your rock across its unglazed underside. What color was the streak?

Yellow pyrite drew a black streak, black limonite drew a yellow streak, and black hematite drew a red streak. (Some minerals, Pough explained, to my mystification, are "not truly black . . . but only look so.") The streaks were brilliant pigments, richer than crayon strokes, deeper than pastel strokes; they were powdery pure pigments bright as greasepaint. It was a wonder the earth wasn't streaked like a Van Gogh landscape, and all the people streaked like warpath Indians.

You performed other testing marvels on your rocks—at least, the people in the books did. They dripped acid on them; they shone ultraviolet lights on them; they split them, sawed them, and set them on fire (diamond "burns easily"). They smelled and tasted them. Cracked arsenic smells like garlic. Epsomite is bitter, halotrichite tastes like ink, soda niter "tastes cooling." Those ardent mineralogists who licked their chrysocolla specimens found that their tongues got stuck.

During these tests, the rocks behaved with scarcely less vigor than the scientists. Borax "swells into great 'worms' as it melts, and finally shrinks to almost nothing." Other minerals "may send up little horns." Some change color when you heat them, or glow, or melt, burn, dissolve, or turn magnetic. Some fly apart (decrepitate). If you should happen to place a hunk of gummite on film, it will take its own picture.

At the end of all these tests, especially if you knew where you found your rocks, you could learn what you had in the paper bags. Or you could, as I did, read the texts' mineral descriptions a thousand times until you hit on something that sounded plausible. You could also go directly to the answers by studying the labeled rocks for sale at the Carnegie Museum shop.

Eventually I identified the rocks. The petrified roses were barite, probably from Oklahoma. The scratchy brown mineral was bauxite—aluminum ore. The black glass was obsidian; the booklet of transparent sheets was mica; the goldeny iridescent handful of soft crystals was chalcopyrite, an ore of copper, whose annoying name I loved to repeat: chalcopyrite. I had shiny green hornblende, rose quartz, starry moss agate, and dull hornfels, which was a mere rock. (A mineral is a pure inorganic compound; you can express its constituents in a chemical formula. A rock is just a mixture of minerals. Worthless, weedy rock is gangue rock.)

I had glassy drops of perlite called Apache tears, bubbly pyrite (fool's gold), and—a favorite—brick-red cinnabar. I had speckly gneiss rock, a chip of crystal tourmaline like a stick of anise candy, and green malachite in a silky chunk. I had milky turquoise, opalized wood, two sorry stalactites, banded jasper, and a lump of coal.

From the book I learned that there was fine stuff hidden in the earth. In the rock underfoot, in the mountain roadside rock, were sealed pockets lined with crystals. You could break a brown rock and find a vug—a pocket—sharp with amethysts.

In Maine, someone with a hammer had discovered a single feldspar crystal twenty feet across. Other New England outcrops have yielded "sparkling blue beryl crystals 18 to 27 feet long." Copper miners find peacock ore, a bronze mineral locked in rock, which tarnishes at once to "an astonishing royal purple" when it hits the light.

The rock I'd seen in my life looked dull because in all ignorance I'd never thought to knock it open. People have cracked ordinary New England pegmatite—big, coarse granite—and laid bare clusters of red garnets, or topaz crystals, chrysoberyl, spodumene, emerald. They held in their hands crystals that had hung in a hole in the dark for a billion years unseen.

I was all for it. I would lay about me right and left with a hammer, and bash the landscape to bits. I would crack the earth's crust like a piñata and spread to the light the vivid prizes in chunks within. Rock collecting was opening the mountains. It was like diving through my own interior blank blackness to remember the startling pieces of a dream: there was a blue lake, a witch, a lighthouse, a yellow path. It was like poking about in a grimy alley and finding an old, old coin. Nothing was as it seemed. The earth was like a shut eye. Mother's not dead, dear—she's only sleeping. Pry open the thin lid and find a crystalline intelligence inside, a rayed and sidereal beauty. Crystals grew inside rock like arithmetical flowers. They lengthened and spread, adding plane to plane in awed and perfect obedience to an absolute geometry that even the stones—maybe only the stones—understood.

The study of minerals reverted alarmingly to the classification of crystals, which in turn smacked dismayingly of math. I was in this, as it were, for my health. Nothing compelled me to finish reading a sentence that began "The macrodomes become, for obvious reasons, *clinodomes*," or "Remember this: *b*-pinacoid equals the brachypinacoid." Yet even the mumbo-jumbo had its charms. It sounded like Sid Caesar. Pough's *Field Guide to Rocks and Minerals* included a diagram of what looked like an angular inner tube or a corduroy hassock. The caption explained the diagram: "Rutile sixling." Here also was a line drawing of a set of penetration twins, and a meticulous side view of a pinacoid, labeled "Side view of pinacoid." Chrysoberyl, I learned—a blue-green aluminum oxide—has the exuberant-sounding "habit" of trilling.

The awesome story of earth's crust's buckling and shifting unfortunately failed to move me in the slightest. But here was an interesting find. Only a quirk of chemistry prevented the ground's being a heap of broken rubble. I hadn't thought of that. Why isn't it all a heap of broken rubble? For the bedrock fractures and cleaves, notoriously; it uplifts, crumbles, splits, shears, and folds. All this action naturally shatters the crust. But it happens that the abundant element silicon is water soluble at high temperatures. This element heals the scars. Dissolved silicon seeps everywhere underground and slips into fissures and veins; it fills in, mends, and cements the rubble, over and over, from age to age. It heals all the thick wounds on the continents' skin and under the oceans; it solidifies as it cools, uplifting, and forms pale veins of scarry quartz running through everything; it dominates the granite bedrock on which we build our cities, the granite interior of mountains, and the beds that underlie the plains.

No one has ever found a rock as old as the earth. All the old rock went under. The age of the earth is 4.8 billion years, and of the oldest rock, a Labrador greenstone, a mere 3.8 billion years. The rock we see is mostly a mishmash of scar tissue and recent rubble. If there were no soluble silicon, how many feet thick, or miles thick, I wondered, would the sterile rubble be?

Many of the rocks in my collection were veins of some bright mineral in a matrix of quartz. The round stones I gathered along Lake Erie's shore were striped with bands of white or tinted quartz. The planet was all healed rubble, rubble joined and smoothed as if a god had rolled it over in his hands like snow.

People who collected rocks called themselves "rockhounds." In the worst of cases, they called their children "pebble pups." Rockhounds seemed to be wild and obsessive amateurs, my kind of people, who had stepped aside from the rush of things to devote themselves to folly.

A collector would be foolish, one book advised, to sell to a gem dealer a fine crystal of, say, ruby or sapphire, when it was obviously of much more value to the collector uncut—a brilliant stub growing from a rough matrix as it was found, a prize specimen in a beloved collection. One book cautioned me against refining any gold I found—any nuggets, dust, or gold-bearing quartz. For I could own or transport all the raw native gold I wanted, but if I refined it in any way I was "obliged by law" to sell it to a licensed gold dealer or the U.S. mint.

These, then, were books which advised, in detail, how to avoid making money, right here in America. Right here in Pittsburgh were people who dug up the nation's mineral wealth, played with it, stored it behind glass, looked at it, fled in a flat-out sprint from anyone who threatened to buy it for dollars, and ultimately gave it to the paper boy. I applauded this with the uncanny, exultant, spreading sensation you have when you realize that your name is really legion—but I wasn't so persnickety that I wasn't inspired by a 1953 story of two Western collectors. These men found two petrified logs that made their Geiger counters click. Uranium-bearing silicates had replaced the logs' wood. The men sold the petrified logs to the Atomic Energy Commission for $35,000.

Some rockhounds had recently taken up scuba diving. These people dove down into "brawling mountain streams" with tanks on their backs to look for crystals underwater, or to pan for gold. The gold panning was especially good under boulders in rapids.

One book included a photograph of a mild-looking hobbyist in his basement workshop: he sawed chunks of Utah wonderstone into wavy, landscapy-looking slabs suitable for wall hangings. Here was a photograph of rockhounds in the field: Two men on a steep desert hillside delightedly smash a flat rock to bits with two hammers. Far below stands a woman in a dress and sensible shoes, doing nothing. Here is their campsite: a sagging black pyramidal tent pitched on the desert floor. A Studebaker fender nudges the foreground. The very hazards of field collecting tempted me: "tramping for miles over rough country," facing cold, heat, rain, cactus, rough lava, insects, rattlesnakes, scorpions, and glaring alkali flats. Collectors fell over boulders and damaged crystals. Their ballpoint pens ran out of ink. They carried sledgehammers, chains, snakebite kits, Geiger counters, canteens, tarps, maps, three-ton hydraulic jacks, mattocks, gold pans, dynamite (see *The Blaster's Handbook,* published by Du Pont), cuff-link boxes, gads, sacks, ultraviolet lamps, pry bars, folding chairs, and the inevitable bathroom tiles.

Getting back home alive only aggravated their problems. If you bring home five hundred pounds of rocks from an average collecting trip, what do you do with them? Splay them attractively about the garden, one book suggested lamely. Give them away. Hold yard sales. One collector left five tons of rough rock in his yard when he moved. The books stopped just short of advising collectors how to deal with their wives.

The problems of storage and display were surprising. A roomful of rocks was evidently as volatile as a roomful of baby raccoons. Once you commit yourself to your charges, you scarcely dare take your eyes off them.

If you have some sky-blue chalcanthite on a shelf, or gypsum, or borax, or trona, it will crumble of its own accord to powder. Your crystals of realgar (an orange-red ore of arsenic) will "disintegrate to a dust of orpiment," which in turn will decompose. Your hanksite and soda niter will absorb water from the air and dissolve into little pools. Your proustite and silver ores will tarnish and then decompose. Your orange beryl will fade to pink, your brown topaz will lose all its color, your polished opals will craze. Finally, your brass-yellow marcasite will release sulfuric acid. The acid will eat your labels, your shelves, and eventually your whole collection.

On the other hand, rock collecting had unique rewards. For example, the thinner you sliced your specimens when you sawed them up, the more specimens you had. In this way you could multiply your collection without leaving home.

When you pry open the landscape, you find wonders—gems made of corpses, even, and excrement. In Puget Sound you could find fossil oysters and clams that had turned to agate, called agatized oysters, agatized clams. In Colorado you could find fossil shrimps turned to scarlet and precious carnelian. People have found dinosaur bones turned to jasper.

Petrified wood is abundant in every county of every state, because soluble silicon seeps everywhere. In Southern states you could find petrified leaves and twigs. There are often worm borings in petrified wood, and inside the opalized tunnels you might find gemmy piles of petrified worm excrement. Dinosaur excrement fossilizes, too. Bird and bat guano petrifies into a mineral called taranakite, which a book described as "unctuous to touch."

I wanted to find these things. I also wanted to find wulfenite crystals the color of cranberries, from Chile, and big transparent cubes of Iceland spar. In Durango, Mexico, I might find rosy adamite, tabular orange sulfenite crystals, mimetite in yellow mammillary crusts, or light green pyramids of scorodite. In Paterson, New Jersey, I could find great pearly crystals of phlogopite, or radiating splinters of white pectolite, or stilbite in bundles like cauliflower. People made extensive mineral discoveries in Tsumeb, Southwest Africa, in Haddam, Connecticut, in Westphalia, Germany, and in Westchester County, New York. Every day for the past two hundred years, one book said, someone has stumbled "on a brand-new outcrop that nobody before has explored." "A lifetime of study," another book said stirringly, "would not make you the master of every phase" of mineralogy.

I thought to specialize in interesting names. In Massachusetts and Connecticut, collectors found a mineral called sillimanite, named for a Yale professor. My specialized collection would feature sillimanite, radio opal, yowah nuts, and agaty potch.

Rock collecting is pleasantly simple for a bookish child, for it scarcely matters which rocks you actually have and which you only imagine having. I might as well throw into my future collection some clams that had turned to agate, too, and some carnelian shrimps. Similarly, field trips don't take much gas. You never think, How could I get a jeep? You forget your condition. You do not see yourself as a figure; you see the world.

From what box canyons have you not extricated yourself, hand over hand, hauling your pickax and Geiger counter in your knapsack? Sharp eyes—you spotted that rattlesnake. I knew you would. You could write a book: the fiery raw glare of the alkali flats, how it shrivels your eyes and blinds your brain in two holes like gimlets, whatever gimlets are, and your jeep tracks crumble over it white on white; and how, when you finally got to the hills you held your hot hammer by its wood handle, planted your big-booted feet in the hillside and felt the hot hill rock through your coarse pants on your thighs. And you broke rocks, hit rocks that cracked like ice in your palm, with your buddy beside you, both of you seeing what you could see. Doubloons, maybe, again, or uranium.

There were now 340 rocks in my room. I labeled them. On November evenings alone in my room while it rained, on February nights while snow piled in the buckeye boughs outside my window, when the paper on the Panama Canal was out of the way, and the Latin memorized, and my friends talked to on the phone until there was nothing left of the schoolday to analyze; while Margaret ran water in the kitchen and Mother talked Molly into going to bed and Father poked the fold from his section of the evening paper and Amy sat silent on the floor with her spelling book and played with the bows on Father's shoes, I played at my maple desk with books and rocks.

From stiff index cards I cut tiny squares, an eighth of an inch by an eighth of an inch. Holding the squares with tweezers, I printed numerals on them, never thinking to print the numerals before cutting the paper. In a dizzying meticulous mess I slid the wee numbers against the drippy rubber slope of a mucilage jar, and wiggled them onto flat faces of the rocks. When the glue dried the next day, I dabbed a brushful of varnish over rock and number. I was cataloguing the collection.

As the books advised, I listed by number in a notebook each specimen's name, date, and locality. "Locality," when it wasn't Mr. Downey's grocery bags, was often the Carnegie Museum shop, where I bought trays of thumbnail minerals glued to cardboard sheets. I soaked the glue from these and labeled them with my own compulsive little numbers. When one day I discovered I had, characteristically, lost the notebook catalogue, I discovered simultaneously that I didn't need it. I knew them by sight: that favorite dry red cinnabar, those Lake Erie ruby granites and flintstones, Mr. Downey's big chunks of pale oolite, dark wavy serpentine, hornblende, gneiss, tourmaline, Apache tears, all of them.

Often I imagined the solitary Mr. Downey, that invisible man from my grandparents' rarefied street of executives and lawyers. Every few years he had unearthed from the cellar his loose khaki pants and his tight hiking boots. He clambered into a pickup truck concealed in his ivy-grown garage, and took off for Oklahoma, where he scouted the lonesome hills with a hammer. He filled his stout sack with moss agate, chalcopyrite, and petrified roses. For reasons unimaginable, he drove back to hushed old Pittsburgh, pulled into his secret driveway, changed his clothes, and sat for years on end at his library desk to study his rocks. When he felt something inside him pulling him down, he called to him the black-haired paper boy, who was collecting quarters on his black bike.

Maybe he hadn't died at all. Maybe he'd simply escaped

underground. He cracked open Pittsburgh like a geode. Who knew what lay inside the streaked hillsides under the high-porched workers' houses, under the streetcar tracks, under the flat or sloping greens of the country-club golf courses, under the dancing-school building, the trout-stocked streams in the highlands or the dried-out stream under the bridge in Frick Park, under the sled-riding hills, the Ellis School, the stained opened cuts on the boulevard roadsides into town—who knew? He had screwed in wedges here and there, tapped them each once or twice, and laid bare the invisible city: crystal-crusted cavities lined with fire opals and red plume agates, where cobalt bloomed and onyx and jacinth grew sharp. He visited the underground corridors where spinel crystals twinned underfoot, and blue cubes of galena cut his hands, and carnelian nodules hung wet overhead among pale octahedrons of fluorite, among frost agates and moonstones, red jasper, blue lazulite, stubs of garnet, black chert. Of course he hadn't come back. Who would?

AFTER I READ *The Field Book of Ponds and Streams* several times, I longed for a microscope. Everybody needed a microscope. Detectives used microscopes, both for the FBI and at Scotland Yard. Although usually I had to save my tiny allowance for things I wanted, that year for Christmas my parents gave me a microscope kit.

In a dark basement corner, on a white enamel table, I set up the microscope kit. I supplied a chair, a lamp, a batch of jars, a candle, and a pile of library books. The microscope kit supplied a blunt black three-speed microscope, a booklet, a scalpel, a dropper, an ingenious device for cutting thin segments of fragile tissue, a pile of clean slides and cover slips, and a dandy array of corked test tubes.

One of the test tubes contained "hay infusion." Hay infusion was a wee brown chip of grass blade. You added water to it, and after a week it became a jungle in a drop, full of one-celled animals. This did not work for me. All I saw in the microscope after a week was a wet chip of dried grass, much enlarged.

Another test tube contained "diatomaceous earth." This was, I believed, an actual pinch of the white cliffs of Dover. On my palm it was an airy, friable chalk. The booklet said it was composed of the silicaceous bodies of diatoms—one-celled creatures that lived in, as it were, small glass jewelry boxes with fitted lids. Diatoms, I read, come in a variety of transparent geometrical shapes. Broken and dead and dug out of geological deposits, they made chalk, and a fine abrasive used in silver polish and toothpaste. What I saw in the microscope must have been the fine abrasive—grit enlarged.

It was years before I saw a recognizable, whole diatom. The kit's diatomaceous earth was a bust.

All that winter I played with the microscope. I prepared slides from things at hand, as the books suggested. I looked at the transparent membrane inside an onion's skin and saw the cells. I looked at a section of cork and saw the cells, and at scrapings from the inside of my cheek, ditto. I looked at my blood and saw not much; I looked at my urine and saw long iridescent crystals, for the drop had dried.

All this was very well, but I wanted to see the wildlife I had read about. I wanted especially to see the famous amoeba, who had eluded me. He was supposed to live in the hay infusion, but I hadn't found him there. He lived outside in warm ponds and streams, too, but I lived in Pittsburgh, and it had been a cold winter.

Finally late that spring I saw an amoeba. The week before, I had gathered puddle water from Frick Park; it had been festering in a jar in the basement. This June night after dinner I figured I had waited long enough. In the basement at my microscope table I spread a scummy drop of Frick Park puddle water on a slide, peeked in, and lo, there was the famous amoeba. He was as blobby and grainy as his picture; I would have known him anywhere.

Before I had watched him at all, I ran upstairs. My parents were still at table, drinking coffee. They, too, could see the famous amoeba. I told them, bursting, that he was all set up, that they should hurry before his water dried. It was the chance of a lifetime.

Father had stretched out his long legs and was tilting back in his chair. Mother sat with her knees crossed, in blue slacks, smoking a Chesterfield. The dessert dishes were still on the table. My sisters were nowhere in evidence. It was a warm evening; the big dining-room windows gave onto blooming rhododendrons.

Mother regarded me warmly. She gave me to understand that she was glad I had found what I had been looking for, but that she and Father were happy to sit with their coffee, and would not be coming down.

She did not say, but I understood at once, that they had

their pursuits (coffee?) and I had mine. She did not say, but I began to understand then, that you do what you do out of your private passion for the thing itself.

I had essentially been handed my own life. In subsequent years my parents would praise my drawings and poems, and supply me with books, art supplies, and sports equipment, and listen to my troubles and enthusiasms, and supervise my hours, and discuss and inform, but they would not get involved with my detective work, nor hear about my reading, nor inquire about my homework or term papers or exams, nor visit the salamanders I caught, nor listen to me play the piano, nor attend my field hockey games, nor fuss over my insect collection with me, or my poetry collection or stamp collection or rock collection. My days and nights were my own to plan and fill.

When I left the dining room that evening and started down the dark basement stairs, I had a life. I sat to my wonderful amoeba, and there he was, rolling his grains more slowly now, extending an arc of his edge for a foot and drawing himself along by that foot, and absorbing it again and rolling on. I gave him some more pond water.

I had hit pay dirt. For all I knew, there were paramecia, too, in that pond water, or daphniae, or stentors, or any of the many other creatures I had read about and never seen: volvox, the spherical algal colony; euglena with its one red eye; the elusive, glassy diatom; hydra, rotifers, water bears, worms. Anything was possible. The sky was the limit.

WHAT DOES IT FEEL LIKE to be alive?

Living, you stand under a waterfall. You leave the sleeping shore deliberately; you shed your dusty clothes, pick your barefoot way over the high, slippery rocks, hold your breath, choose your footing, and step into the waterfall. The hard water pelts your skull, bangs in bits on your shoulders and arms. The strong water dashes down beside you and you feel it along your calves and thighs rising roughly back up, up to the roiling surface, full of bubbles that slide up your skin or break on you at full speed. Can you breathe here? Here where the force is greatest and only the strength of your neck holds the river out of your face? Yes, you can breathe even here. You could learn to live like this. And you can, if you concentrate, even look out at the peaceful far bank where maples grow straight and their leaves lean down. For a joke you try to raise your arms. What a racket in your ears, what a scattershot pummeling!

It is time pounding at you, time. Knowing you are alive is watching on every side your generation's short time falling away as fast as rivers drop through air, and feeling it hit.

Who turned on the lights? You did, by waking up: you flipped the light switch, started up the wind machine, kicked on the flywheel that spins the years. Can you catch hold of a treetop, or will you fly off the diving planet as she rolls? Can you ride out the big blow on a coconut palm's trunk until you fall asleep again, and the winds let up? You fall asleep again, and you slide in a dream to the palm tree's base; the winds die off, the lights dim, the years slip away as you idle there till you die in your sleep, till death sets you cruising down the Tamiami Trail.

Knowing you are alive is feeling the planet buck under you, rear, kick, and try to throw you; you hang on to the ring. It is riding the planet like a log downstream, whooping. Or, conversely, you step aside from the dreaming fast loud routine and feel time as a stillness about you, and hear the silent air asking in so thin a voice, Have you noticed yet that you will die? Do you remember, remember, remember? Then you feel your life as a weekend, a weekend you cannot extend, a weekend in the country.

O Augenblick verweile.

My friend Judy Schoyer was a thin, messy, shy girl whose thick blond curls lapped over her glasses. Her cheeks, chin, nose, and blue eyes were round; the lenses and frames of her glasses were round, and so were her heavy curls. Her long spine was supple; her legs were long and thin so her knee socks fell down. She did not care if her knee socks fell down. When I first knew her, as my classmate at the Ellis School, she sometimes forgot to comb her hair. She was so shy she tended not to move her head, but only let her eyes rove about. If my mother addressed her, or a teacher, she held her long-legged posture lightly, alert, like a fawn ready to bolt but hoping its camouflage will work a little longer.

Judy's family were members of the oldest, most liberal, and best-educated ranks of Pittsburgh society. They were Unitarians. I visited her Unitarian Sunday school once. There we folded paper to make little geese; it shocked me to the core. One of her linear ancestors, Edward Holyoke, had been president of Harvard University in the eighteenth century, which fact paled locally before the greater one, that her great-grandfather's brother had been one of the founding members of Pittsburgh's Duquesne Club. She was related also to Pittsburgh's own Stephen Foster.

Judy and her family passed some long weekends at a family farmhouse in the country on a little river, the nearest town to which was Paw Paw, West Virginia. When they were going to the farm, they said they were going to Paw

Paw. The trip was a four-hour drive from Pittsburgh. Often they invited me along.

There in Paw Paw for the weekend I imagined myself in the distant future remembering myself now, twelve years old with Judy. We stood on the high swinging plank bridge over the river, in early spring, watching the first hatch of small flies hover below us.

The river was a tributary of the distant Potomac—a tributary so stony, level, and shallow that Judy's grandmother regularly drove her old Model A Ford right through it, while we hung out over the running boards to try and get wet. From above the river, from the hanging center of the swinging bridge, we could see the forested hill where the big house stood. There at the big house we would have dinner, and later look at the Gibson girls in the wide, smelly old books in the cavernous living room, only recently and erratically electrified.

And from the high swinging bridge we could see in the other direction the log cabin, many fields away from the big house, where we children stayed alone: Judy and I, and sometimes our friend Margaret, who had a dramatic, somewhat morbid flair and who wrote poetry, and Judy's good-natured younger brother. We cooked pancakes in the cabin's fireplace; we drew water in a bucket from the well outside the door.

By Friday night when we'd carried our duffel and groceries from the black Model A at the foot of the hill, or over the undulating bridge if the river was high, when we children had banged open the heavy log-cabin door, smelled the old logs and wood dust, found matches and lighted the kerosene lanterns, and in the dark outside had drawn ourselves a bucket of sweet water (feeling the rope go slack and hearing the bucket hit, then feeling the rope pull as the bucket tipped and filled), and hunted up wood for a fire, smelled the loamy nighttime forest again, and heard the whippoorwill—by that time on Friday night I was already grieving and mourning, only just unpacking my nightgown, because here it was practically Sunday afternoon and time to go.

"What you kids need," Mrs. Schoyer used to say, "is more exercise."

How exhilarating, how frightening, to ride the tippy Model A over the shallow river to the farm at Paw Paw, to greet again in a new season the swaying bridge, the bare hills, the woods behind the log cabin, the hayloft in the barn—and know I had just so many hours. From the minute I set foot on that land across the river, I started ticking like a timer, fizzing like a fuse.

On Friday night in the log cabin at Paw Paw I watched the wild firelight on Judy's face as she laughed at something her cheerful brother said, laughed shyly even here. When she laughed, her cheeks rose and formed spheres. I loved her spherical cheeks and knocked myself out to make her laugh. I could hardly see her laughing eyes behind her glasses under hanging clumps of dark-blond curls. She was nimble, sway-backed, long-limbed, and languid as a heron, and as abrupt. In Pittsburgh she couldn't catch a ball—nearsighted; she perished of bashfulness at school sports. Here she could climb a tree after a kitten as smoothly as a squirrel could, and run down her nasty kicking pony with authority, and actually hit it, and scoop up running hens with both swift arms. She spoke softly and not often.

Judy treated me with amused tolerance. At school I was, if not a central personage, at least a conspicuous one; and I had boyfriends all along and got invited to the boys' school dances. Nevertheless, Judy put up with me, not I with Judy. She possessed a few qualities that, although they counted for nothing at school, counted, I had to admit, with me. Her goodness was both intrinsic and a held principle. This thin, almost speechless child had moral courage. She intended her own life—starting when she was about ten—to be not only harmless but good. I considered Judy's goodness, like Judy's farm, a nice place to visit. She put up with my fast-talking avoidance of anything that smacked of manual labor. That she was indulging me altogether became gradually clear to both of us—though I pretended I didn't know it, and Judy played along.

On Saturday mornings in Paw Paw we set out through the dewy fields. I could barely lay one foot before the other along the cowpath through the pasture, I was so nostalgic for this scene already, this day just begun, when Judy and I were twelve. With Margaret we boiled and ate blue river mussels; we wrote and staged a spidery melodrama. We tried to ride the wretched untrained pony, which scraped us off under trees. We chopped down a sassafras tree and made a dirty tea; and we started to clean a run-over snake, in order to make an Indian necklace from its delicate spine, but it smelled so bad we quit.

After Saturday-night dinner in the big-house dining room—its windows gave out on the cliffside treetops—Mr. Schoyer told us, in his calm, ironic voice, Victor Hugo's story of a French sailor who was commended for having heroically captured a cannon loose on the warship's deck, and then hanged for having loosed the cannon in the first place. There were usually a dozen or more of us around the table, rapt. When the household needed our help, Mrs. Schoyer made mild, wry suggestions, almost diffidently.

I would have liked going to prison with the Schoyers. My own family I loved with all my heart; the Schoyers fascinated me. They were not sharply witty but steadily wry. In Pittsburgh they invited foreigners to dinner. They went to art galleries, they heard the Pittsburgh Symphony. They weren't tan. Mr. Schoyer, who was a corporation lawyer, had majored in classical history and literature at Harvard. Like my father, he had studied something that had no direct bearing on the clatter of coin. He was always the bemused scholar, mild and democratic, posing us children friendly questions as if Pittsburgh or Paw Paw were Athens and he fully expected to drag from our infant brains the Pythagorean theorem. What do you make of our new President? Your position on capital punishment? Or, conversationally, after I had been branded as a lover of literature, "You recall that speech of Pericles, don't you?" or "Won't you join me in reading 'A Shropshire Lad' or 'Ballad of East and West'?" At Paw Paw the Schoyers did every wholesome thing but sing. None of them could carry a tune.

If there was no moon that night, we children took a flashlight down the steep dirt driveway from the big house and across the silvery pastures to the edge of the woods where the log cabin stood. The log cabin stayed empty, behind an old vine-hung gate, except when we came. In front of the cabin we drew water from the round stone well; under the cabin we put milk and butter in the cold cellar, which was only a space dug in the damp black dirt—dirt against which the butter's wrap looked too thin.

That was the farm at Paw Paw, West Virginia. The farm lay far from the nearest highway, off three miles of dirt road. When at the end of the long darkening journey from Pittsburgh we turned down the dirt road at last, the Schoyers' golden retriever not unreasonably began to cry, and so, unreasonably, invisibly, did I. Some years when the Schoyers asked me to join them I declined miserably, refused in a swivet, because I couldn't tolerate it, I loved the place so.

I knew what I was doing at Paw Paw: I was beginning the lifelong task of tuning my own gauges. I was there to brace myself for leaving. I was having my childhood. But I was haunting it, as well, practically reading it, and preventing it. How much noticing could I permit myself without driving myself round the bend? Too much noticing and I was too self-conscious to live; I trapped and paralyzed myself, and dragged my friends down with me, so we couldn't meet each other's eyes, my own loud awareness damning us both. Too little noticing, though—I would risk much to avoid this—and I would miss the whole show. I would wake on my deathbed and say, What was that?

YOUNG CHILDREN HAVE NO SENSE OF WONDER. They bewilder well, but few things surprise them. All of it is new to young children, after all, and equally gratuitous. Their parents pause at the unnecessary beauty of an ice storm coating the trees; the children look for something to throw. The children who tape colorful fall leaves to the schoolroom windows and walls are humoring the teacher. The busy teacher halts on her walk to school and stoops to pick up fine bright leaves "to show the children"—but it is she, now in her sixties, who is increasingly stunned by the leaves, their brightness all so much trash that litters the gutter.

This year at the Ellis School my sister Amy was in the fifth grade, with Mrs. McVicker. I remembered Mrs. McVicker fondly. Every year she reiterated the familiar (and, without a description of their mechanisms, the sentimental) mysteries that schoolchildren hear so often and so indifferently: that each snowflake is different, that some birds fly long distances, that acorns grow into oaks. Caterpillars turn into butterflies. The stars are large and very far away. She struck herself like a gong with these same mallets every year—a sweet old schoolteacher whom we in our time had loved and tolerated for her innocence.

Now that I was an aging veteran of thirteen or so, I was becoming case-softened myself. Imperceptibly I had shed my indifference. I was getting positively old: the hatching of wet robins in the spring moved me. I saw them from the school library window, as if on an educational film: a robin sprawled on a nest in the oak, and four miserable hatchlings appeared. They peeped. I knew this whole story; who didn't?

Nevertheless I took to checking on the robins a few times a day. Their mother rammed worms and bugs down their throats; they grew feathers and began to hop up and down in the nest. Bit by bit they flew away; I saw them from the schoolyard taking test flights under the oak. Glory be, I thought during all those weeks, hallelujah, and never told a soul.

Even my friends began to seem to me marvelous: Judy Schoyer laughing shyly, her round eyes closed, and quick Ellin Hahn, black-haired and ruddy, who bestrode the social world like a Colossus, saying always just the right and funny thing. Where had these diverse people come from, really? I watched little Molly turn from a baby into a child and become not changed so much as ever more herself, kind-hearted, nervous, both witty and humorous: was this true only in retrospect? People's being themselves, year after year, so powerfully and so obliviously—what was it? Why was it so appealing? Personality, like beauty, was a mystery; like beauty, it was useless. These useless things were not, however, flourishes and embellishments to our life here, but that life's center; they were its truest note, the heart of its form, which drew back our thoughts repeatedly.

Somewhere between one book and another a child's passive acceptance had slipped away from me also. I could no longer see the world's array as a backdrop to my private play, a dull, neutral backdrop about which I had learned all I needed to know. I had been chipping at the world idly, and had by accident uncovered vast and labyrinthine further worlds within it. I peered in one day, stepped in the next, and soon wandered in deep over my head. Month after month, year after year, the true and brilliant light, and the complex and multifaceted coloration, of this actual, historical, waking world invigorated me. Its vastness extended everywhere I looked, and precisely where I looked, just as forms grew under my gaze as I drew.

This was the enthusiasm of a child, like that of a field-working scientist, and like that of the artist making a pencil study. One took note; one took notes. The subject of the study was the world's things: things to sort into physical categories, and things to break down into physical structures.

I was not to discover literature and ideas for a few more years. All I had awakened to was the world's wealth of information. I was reading books on drawing, painting, rocks, criminology, birds, moths, beetles, stamps, ponds and streams, medicine. (Somehow I missed those other childhood mainstays, astronomy, coins, and dinosaurs.) How I wished I could find agreeable books on thin air! For everything, I had gathered, was something. And for me, during those few years before I vanished into a blinded rage, everything was interesting.

Nothing could be less apparently interesting, for example, than a certain infuriatingly dull sight I always looked at with hatred. It was raining and Mother was driving us along one of Pittsburgh's clogged narrow highways. I looked out through the rain on the window and saw by the roadside the raw cuts the road builders carved through the rolling rocky hills, carved long dreary decades ago, to lay the road. Blasting bores scarred these banks of sandstone and shale in streaks; gritty rain streamed down their cut faces and dissolved the black soot and coal dust and car exhaust. The car stopped and started. I stared dully through the spotted windshield. Gray rivulets poured down the rock, mile after highway mile, and puddled at the berm where the rock met the winter-killed grass and mud.

This sight slew me in my seat. It was so dull it unstrung me, so I could barely breathe. How could I flee it, the very landscape, the dull rock, the bleak miles, the dark rain? I slumped under the weight of my own passive helplessness. Sometimes I memorized billboards. I tried traveling with my eyes closed, and that was even worse.

But now I knew that even rock was interesting—at least in theory. Mr. Pough and Herr Mohs could stand here mightily in the rain, singing songs and swinging picks into the rock cuts by the side of the road. Even I could tap some shale just right, rain or shine, and open the rock to bones of fossil fish. There might be trilobites on the hilltops, star sapphires. Right along these wretched rainy roads, Mohs and Pough could have, as the saying went, a field day.

If even rock was interesting, if even this ugliness was worth whole shelves at the library, required sophisticated tools to study, and inspired grown men to crack mountains and saw crystals—then what wasn't?

Everything in the world, every baby, city, tetanus shot, tennis ball, and pebble, was an outcrop of some vast and hitherto concealed vein of knowledge, apparently, that had compelled people's emotions and engaged their minds in the minutest detail without anyone's having done with it. There must be bands of enthusiasts for everything on earth—fanatics who shared a vocabulary, a batch of technical skills and equipment, and, perhaps, a vision of some single slice of the beauty and mystery of things, of their complexity, fascination, and unexpectedness. There was no one here but us fanatics: bird-watchers, infielders, detectives, poets, rock collectors, and, I inferred, specialists in things I had not looked into—violin makers, fishermen, Islamic scholars, opera composers, people who studied Bali, vials of air, bats. It seemed to take all these people working full time to extract the interest from everything and articulate it for the rest of us.

Every least thing I picked up was proving to be the hanging end of a very long rope.

For the sentimental Mrs. McVicker I had written on assignment a paper on William Gorgas—the doctor in charge of workers' health during the digging of the Panama Canal. Liking that, I wrote another, on Walter Reed. The struggle against yellow fever fired me, and I retained an interest in medicine, especially epidemiology. So now, a few years later, on the couch on the sunporch, I was reading Paul de Kruif's overwrought *Microbe Hunters*.

Old Anton Leeuwenhoek looked through his lenses at a drop of rainwater and shouted to his daughter, "Come here! Hurry! There are little animals in this rainwater! . . . They swim! They play around!" His microscope "showed little things to him with a fantastic clear enormousness." My microscope was similar. Since I had found the amoeba, I regularly found little animals. I found them in rainwater. I let a bowl of rainwater sit by the basement furnace for a week. When I examined a drop at low power, sure enough, little animals swam, and played around, with fantastic clear enormousness.

Not only was the roadside rock interesting; even the rainwater that streamed down its cut face was interesting. Mineral crystals made the rock; lively animals made the rain. Now when I traveled the grim highways and saw the dull rock receive the dull rain, and realized there would be nothing else to look at until we got where we were going, and Mother and I were all talked out—now when I felt the familiar restless hatred begin to rise at the stupidity and ugliness of this sight, I bade myself look directly at some streaky rock cut and said to myself, thundered to myself, "Think!"

Everywhere, things snagged me. The visible world turned me curious to books; the books propelled me reeling back to the world.

At school I saw a searing sight. It turned me to books; it turned me to jelly; it turned me much later, I suppose, into an early version of a runaway, a scapegrace. It was only a freshly hatched Polyphemus moth crippled because its mason jar was too small.

The mason jar sat on the teacher's desk; the big moth emerged inside it. The moth had clawed a hole in its hot cocoon and crawled out, as if agonizingly, over the course of an hour, one leg at a time; we children watched around the desk, transfixed. After it emerged, the wet, mashed thing turned around walking on the green jar's bottom, then painstakingly climbed the twig with which the jar was furnished.

There, at the twig's top, the moth shook its sodden clumps of wings. When it spread those wings—those beautiful wings—blood would fill their veins, and the birth fluids on the wings' frail sheets would harden to make them tough as sails. But the moth could not spread its wide wings at all; the jar was too small. The wings could not fill, so they hardened while they were still crumpled from the cocoon. A smaller moth could have spread its wings to their utmost in that mason jar, but the Polyphemus moth was big. Its gold furred body was almost as big as a mouse. Its brown, yellow, pink, and blue wings would have extended six inches from tip to tip, if there had been no mason jar. It would have been big as a wren.

The teacher let the deformed creature go. We all left the classroom and paraded outside behind the teacher with pomp and circumstance. She bounced the moth from its jar and set it on the school's asphalt driveway. The moth set out walking. It could only heave the golden wrinkly clumps where its wings should have been; it could only crawl down the school driveway on its six frail legs. The moth crawled down the driveway toward the rest of Shadyside, an area of fine houses, expensive apartments, and fashionable shops. It crawled down the driveway because its shriveled wings were glued shut. It crawled down the driveway toward Shadyside, one of several sections of town where people like me were expected to settle after college, renting an apartment until they married one of the boys and bought a house. I watched it go.

I knew that this particular moth, the big walking moth, could not travel more than a few more yards before a bird or a cat began to eat it, or a car ran over it. Nevertheless, it was crawling with what seemed wonderful vigor, as if, I thought at the time, it was still excited from being born. I watched it go till the bell rang and I had to go in. I have told this story before, and may yet tell it again, to lay the moth's ghost, for I still see it crawl down the broad black driveway, and I still see its golden wing clumps heave.

I had not suspected, among other things, that moths came so big. From a school library book I learned there were several such enormous American moths, all wild silk moths which spun cocoons, and all common.

Gene Stratton Porter's old *Moths of the Limberlost* caught my eye; for some years after I read it, it was my favorite book. From one of its queer painted photographs I learned what the Polyphemus moth would have looked like whole: it was an unexpected sort of beauty, brown and wild. It had pink stripes, lavender crescents, yellow ovals—all sorts of odd colors no one would think to combine. Enormous blue eye-spots stared eerily from its hind wings. Coincidentally, it was in the Polyphemus chapter that the book explained how a hatched moth must spread its wings quickly, and fill them with blood slowly, before it can fly.

Gene Stratton Porter had been a vigorous, loving kid who grew up long ago near a swampy wilderness of Indiana, and had worked up a whole memorable childhood out of insects, of all things, which I had never even noticed, and my childhood was half over.

When she was just a tot, she learned how entomologists carry living moths and butterflies without damaging them. She commonly carried a moth or butterfly home from her forest and swamp wanderings by lightly compressing its thorax between thumb and index finger. The insect stops moving but is not hurt; when you let it go, it flies away.

One day, after years of searching, she found a yellow swallowtail. This is not the common tiger swallowtail butterfly, but *Papilio turnus:* "the largest, most beautiful butterfly I had ever seen." She held it carefully in the air, its wings high over the back of her fingers. She wanted to show the fragile, rare creature to her father and then carry it back to precisely where she found it. But she was only a child, and so she came running home with it instead of walking. She tripped, and her fingers pinched through the butterfly's thorax. She broke it to pieces. And that was that. It was like one of Father's bar jokes.

There was a terror connected with moths that attracted and repelled me. I would face down the terror. I continued reading about moths, and branched out to other insects.

I liked the weird horned beetles rumbling along everywhere, even at the country club, whose names were stag, elephant, rhinoceros. They were so big I could hear them walk; their sharp legs scraped along the poolside concrete. I liked the comical true bugs, like the red-and-blue-striped leafhoppers, whose legs looked like yellow plastic; they hopped on roses in the garden at home. At Lake Erie I watched the solitary wasps that hunted along the beach path; they buried their paralyzed caterpillar prey in holes they dug so vigorously the sand flew. I even liked the dull little two-winged insects, the diptera, because this order contained mosquitoes, about several species of which I knew something because they bore interesting diseases. I studied under the microscope our local mosquitoes in various stages—a hairy lot—dipped in a cup from Molly's wading pool.

To collect insects I equipped myself with the usual paraphernalia: glass-headed pins, a net, and a killing jar. It was insects in jars again—but unlike the hapless teacher who put the big moth's cocoon in the little mason jar, I knew, I thought, what I was doing. In the bottom of the killing jar—formerly a pickle jar—I laid a wad of cotton soaked in cleaning fluid containing carbon tetrachloride, which compound I thrilled myself by calling, offhandedly, "carbon tet." A circle of old door screen prevented the insects' tangling in the cotton. I placed each insect on the screen and quickly tightened the jar lid. Then, as if sensitively, I looked away. After a suitable interval I poured out the dead thing as carefully as I could, and pinned it and its festive, bunting-like row of fluttering labels in a cigar box. My grandfather had saved the cigar boxes, one for each order of insect; they smelled both sharp and sweet, of cedar and leaf tobacco. I pinned the insects in rows, carefully driving the pins through chitinous thoraxes just where the books indicated. Four beetles I collected were so big they had a cigar box to themselves.

Once I returned to my attic bedroom after four weeks at summer camp. There, beside the detective table, under the plaster-stain ship, was the insect collection, a stack of cigar boxes. I checked the boxes. In the big beetles' cigar box I found a rhinoceros beetle crawling on its pin. The pin entered the beetle through that triangle in the thorax between the wing-cover tops; it emerged ventrally above and between the legs. The big black beetle's six legs hung down waving in the air, well above the floor of the cigar box. It crawled and never got anywhere. It must have been pretty dehydrated; the attic was hot. Presumably the beetle's legs had been waving in the air like that in search of a footing for the past four weeks.

I hated insects; that was the fact. I never caught my stamp collection trying to crawl away.

Butterflies die with folded wings. Before they're mounted, butterflies require an elaborate chemical treatment to relax their dead muscles, a bit more every day, so you can spread their brittle wings without shattering them. After a few grueling starts at this relaxing and spreading of dead butterflies, I avoided it. When on rare occasions I killed butterflies, I stuck them away somewhere and forgot about them.

One hot evening I settled on my bed in my summer nightgown with a novel I had looked forward to reading. I lay back, opened the book, and a dead butterfly dropped headfirst on my bare neck. I jumped up, my skin crawling, and it slid down my nightgown. Somehow it stuck to my sweaty skin; when I brushed at it—whooping aloud—it fragmented, and pieces stuck to my hands and rained down on the floor. Most of the dead butterfly, which still looked as if it were demurely praying while falling apart, with folded yellow wings in shreds and a blasted black body, fell out on my foot. I brushed broken antennae and snapped legs from my neck; I wiped a glittering yellow dust of wing scales from my belly, and they stuck to my palm.

I hated insects; that I knew. Fingering insects was touching the rim of nightmare. But you have to study something. I never considered turning away from them just because I was afraid of them.

I liked their invisibility; they did not matter, so they did not exist. People's nervous systems edited out the sight of insects before it reached their brains; my seeing insects let me live alongside human society in a different sensory world, just as insects themselves do. That I collected specimens at the country-club pool pleased me; I did not really mind that my friends turned bilious when I showed them my prizes. I loved the sport of catching butterflies; they took bad hops, like aerial grounders. (I did not know then that the truly athletic, life-loving entomologists study dragonflies, which are fantastically difficult to catch—fast, sharp-eyed, hard to outwit.) Cringing, I taught myself to paralyze butterflies through the net, holding them lightly at the thorax as Gene Stratton Porter had done. I brought them out of the net and let them fly away—lest they fall on me dead later.

How confidently I had overlooked all this—rocks, bugs, rain. What else was I missing?

I opened books like jars. Here between my hands, here between some book's front and back covers, whose corners poked dents in my palm, was another map to the neighborhood I had explored all my life, and fancied I knew, a map depicting hitherto invisible landmarks. After I learned to see those, I looked around for something else. I never knew where my next revelation was coming from, but I knew it was coming—some hairpin curve, some stray bit of romance or information that would turn my life around in a twinkling.

I INTENDED TO LIVE the way the microbe hunters lived. I wanted to work. Hard work on an enormous scale was the microbe hunters' stock-in-trade. They took a few clear, time-consuming steps and solved everything. In those early days of germ theory, large disease-causing organisms, whose cycles traced straightforward patterns, yielded and fell to simple procedures. I would know just what to do. I would seize on the most casual remarks of untutored milkmaids. When an untutored milkmaid remarked to me casually, "Oh, everyone knows you won't get the smallpox if you've had the cowpox," I would perk right up.

Microbe Hunters sent me to a biography of Louis Pasteur. Pasteur's was the most enviable life I had yet encountered. It was his privilege to do things until they were done. He established the germ theory of disease; he demonstrated convincingly that yeasts ferment beer; he discovered how to preserve wine; he isolated the bacillus in a disease of silkworms; he demonstrated the etiology of anthrax and produced a vaccine for it; he halted an epidemic of cholera in fowls and inoculated a boy for hydrophobia. Toward the end of his life, in a rare idle moment, he chanced to read some of his early published papers and exclaimed (someone overheard), "How beautiful! And to think that I did it all!" The tone of this exclamation was, it seemed to me, astonished and modest, for he had genuinely forgotten, moving on.

Pasteur had not used up all the good work. Mother told me again and again about one of her heroes, a doctor working for a federal agency who solved a problem that arose in the late forties. Premature babies, and only premature babies, were turning up blind, in enormous numbers. Why? What do premature babies have in common?

"Look in the incubators!" Mother would holler, and knock the side of her head with the heel of her hand, holler outraged, glaring far behind my head as she was telling me this story, holler, "Look in the incubators!" as if at her wit's end facing a roomful of doctors who wrung their useless hands and accepted this blindness as one of life's tough facts. Mother's hero, like all of Mother's heroes, accepted nothing. She rolled up her sleeves, looked in the incubators, and decided to see what happened if she reduced the oxygen in the incubator air. That worked. Too much oxygen had been blinding them. Now the babies thrived; they got enough oxygen, and they weren't blinded. Hospitals all over the world changed the air mixture for incubators, and prematurity no longer carried a special risk of blindness.

Mother liked this story, and told it to us fairly often. Once she posed it as a challenge to Amy. We were all in the living room, waiting for dinner. "What would you do if you noticed that all over the United States, premature babies were blind?" Without even looking up from her homework, Amy said, "Look in the incubators. Maybe there's something wrong in the incubators." Mother started to whoop for joy before she realized she'd been had.

Problems still yielded to effort. Only a few years ago, to the wide-eyed attention of the world, we had seen the epidemic of poliomyelitis crushed in a twinkling, right here in Pittsburgh.

We had all been caught up in the polio epidemic: the early neighbor boy who wore one tall shoe, to which his despairing father added another two soles every year; the girl in the iron lung reading her schoolbook in an elaborate series of mirrors while a volunteer waited to turn the page; my friend who limped, my friend who rolled everywhere in a wheelchair, my friend whose arm hung down, Mother's friend who walked with crutches. My beloved dressed-up aunt, Mother's sister, had come to visit one day and, while she was saying hello, flung herself on the couch in tears; her son had it. Just a touch, they said, but who could believe it?

When Amy and I had asked, Why do we have to go to bed so early? Why do we have to wash our hands again? we knew Mother would kneel to look us in the eyes and answer in a low, urgent voice, So you do not get polio. We heard polio discussed once or twice a day for several years.

And we had all been caught up in its prevention, in the wild ferment of the early days of the Salk vaccine, the vaccine about which Pittsburgh talked so much, and so joyously, you could probably have heard the crowd noise on the moon.

In 1953, Jonas Salk's Virus Research Laboratory at the University of Pittsburgh had produced a controversial vaccine for polio. The small stories in the Pittsburgh *Press* and the *Post-Gazette* were coming out in *Life* and *Time*. It was too quick, said medical colleagues nationwide: Salk had gone public without first publishing everything in the journals. He rushed out a killed-virus serum without waiting for a safe live-virus one, which would probably be better. Doctors walked out of professional meetings; some quit the foundation that funded the testing. Salk was after personal glory, they said. Salk was after money, they said. Salk was after big prizes.

Salk tested the serum on five thousand Pittsburgh schoolchildren, of whom I was three, because I kept changing elementary schools. Our parents, like ninety-five percent of all Pittsburgh parents, signed the consent forms. Did the other mothers then bend over the desk in relief and sob? I don't know. But I don't suppose any of them gave much of a damn what Salk had been after.

When Pasteur died, near a place wonderfully called Saint-Cloud, he murmured to the devoted assistants who surrounded his bed, "*Il faut travailler.*"

Il faut indeed *travailler*—no one who grew up in Pittsburgh could doubt it. And no one who grew up in Pittsburgh could doubt that the great work was ongoing. We breathed in optimism—not coal dust—with every breath. What couldn't be done with good hard *travail*?

The air in Pittsburgh had been dirty; now we could see it was clean. An enormous, pioneering urban renewal was under way; the newspapers pictured fantastic plans, airy artists' watercolors, which we soon saw laid out and built up in steel and glass downtown. The Republican Richard King Mellon had approached Pittsburgh's Democratic, Catholic mayor, David L. Lawrence, and together with a dozen business leaders they were razing the old grim city and building a sparkling new one; they were washing the very air. The Russians had shot Sputnik into outer space. In Shippingport, just a few miles down the Ohio River, people were building a generating plant that used atomic energy—an idea that seemed completely dreamy, but there it was. A physicist from Bell Laboratories spoke to us at school about lasers; he was about as wrought up a man as I had ever seen. You could not reasonably believe a word he said, but you could see that he believed it.

We knew that "Doctor Salk" had spent many years and many dollars to produce the vaccine. He commonly worked sixteen-hour days, six days a week. Of course. In other laboratories around the world, other researchers were working just as hard, as hard as Salk and Pasteur. Hard work bore fruit. This is what we learned growing up in Pittsburgh, growing up in the United States.

Salk had isolated seventy-four strains of polio virus. It took him three years to verify the proposition that a workable vaccine would need samples of only three of these strains. He grew the virus in tissues cultured from monkey kidneys. The best broth for growing the monkey tissue proved to be Medium Number 199; it contained sixty-two ingredients in careful proportion.

This was life itself: the big task. Nothing exhilarated me more than the idea of a life dedicated to a monumental worthwhile task. Doctor Salk never watched it rain and wished he had never been born. How many shovelfuls of dirt did men move to dig the Panama Canal? Two hundred and forty million cubic yards. It took ten years and twenty-one thousand lives and $336,650,000, but it was possible.

I thought a great deal about the Panama Canal, and

always contemplated the same notion: You could take more time, and do it with teaspoons. I saw myself and a few Indian and Caribbean co-workers wielding teaspoons from our kitchen: Towle, Rambling Rose. And our grandchildren, and their grandchildren. Digging the canal across the isthmus at Panama would tear through a good many silver spoons. But it could be done, in theory and therefore in fact. It was like Mount Rushmore, or Grand Coulee Dam. You hacked away at the landscape and made something, or you did not do anything, and just died.

How many filaments had Thomas Edison tried, over how many years, before he found one workable for incandescence? How many days and nights over how many years had Marie Curie labored in a freezing shed to isolate radium? I read a biography of George Washington Carver: so many years on the soybean, the peanut, the sweet potato, the waste from ginning cotton. I read biographies of Abraham Lincoln, Thomas Edison, Daniel Boone.

It was all the same story. You have a great idea and spend grinding years at dull tasks, still charged by your vision. All the people about whom biographies were not written were people who failed to find something that took years to do. People could count the grains of sand. In my own life, as a sideline, and for starters, I would learn all the world's languages.

What if people said it could not be done? So much the better. We grew up with the myth of the French Impressionist painters, and its queer implication that rejection and ridicule guaranteed, or at any rate signaled, a project's worth. When little George Westinghouse at last figured out how to make air brakes, Cornelius Vanderbilt of the New York Central Railroad said to him, "Do you mean to tell me with a straight face that a moving train can be stopped with wind?" "They laughed at Orville," Mother used to say when someone tried to talk her out of a wild scheme, "and they laughed at Wilbur."

I had small experience of the evil hopelessness, pain, starvation, and terror that the world spread about; I had barely seen people's malice and greed. I believed that in civilized countries, torture had ended with the Enlightenment. Of nations' cruel options I knew nothing. My optimism was endless; it grew sky-high within the narrow bounds of my isolationism. Because I was all untried courage, I could not allow that the loss of courage was a real factor to be reckoned in. I put my faith in willpower, that weak notion by which children seek to replace the loving devotion that comes from intimate and dedicated knowledge. I believed that I could resist aging by willpower.

I believed then, too, that I would never harm anyone. I usually believed I would never meet a problem I could not solve. I would overcome any weakness, any despair, any fear. Hadn't I overcome my fear of the ghosty oblong that coursed round my room, simply by thinking it through? Everything was simple. You found good work, learned all about it, and did it.

Questions of how to act were also transparent to reason. Right and wrong were easy to discern: I was right, and Amy was wrong. Many of my classmates stole things, but I did not. Sometimes, in a very tight spot, when at last I noticed I had a moral question on my hands, I asked myself, What would Christ have done? I had picked up this method (very much on the sly—we were not supposed really to believe these things) from Presbyterian Sunday school, from summer camp, or from any of the innumerable righteous orange-bound biographies I read. I had not known it to fail in the two times I had applied it.

As for loss, as for parting, as for bidding farewell, so long, thanks, to love or a land or a time—what did I know of parting, of grieving, mourning, loss? Well, I knew one thing; I had known it all along. I knew it was the kicker. I knew life pulled you in two; you never healed. Mother's emotions ran high, and she suffered sometimes from a web of terrors, because, she said, her father died when she was seven; she still missed him.

My parents played the Cole Porter song "It's All Right with Me." When Ella Fitzgerald sang, "There's someone I'm trying so hard to forget—don't you want to forget someone too?," these facile, breathy lyrics struck me as an unexpectedly true expression of how it felt to be alive. This was experience at its most private and inarticulate: longing and loss. "It's the wrong time, it's the wrong place, though your face is charming, it's the wrong face." I was a thirteen-year-old child; I had no one to miss, had lost no one. Yet I suspect most children feel this way, probably all children feel this way, as adults do; they mourn this absence or loss of someone, and sense that unnamable loss as a hole or hollow moving beside them in the air.

Loss came around with the seasons, blew into the house when you opened the windows, piled up in the bottom desk and dresser drawers, accumulated in the back of closets, heaped in the basement starting by the furnace, and came creeping up the basement stairs. Loss grew as you did, without your consent; your losses mounted beside you like earthworm castings. No willpower could prevent someone's dying. And no willpower could restore someone dead, breathe life into that frame and set it going again in the room with you to meet your eyes. That was the fact of it. The strongest men and women who had ever lived had presumably tried to resist their own deaths, and now they were dead. It was on this fact that all the stirring biographies coincided, concurred, and culminated.

Time itself bent you and cracked you on its wheel. We were getting ready to move again. I knew I could not forever keep riding my bike backward into ever-older neighborhoods to look the ever-older houses in the face. I tried to memorize the layout of this Richland Lane house, but I couldn't force it into my mind while it was still in my bones.

I saw already that I could not in good faith renew the increasingly desperate series of vows by which I had always tried to direct my life. I had vowed to love Walter Milligan forever; now I could recall neither his face nor my feeling, but only this quondam urgent vow. I had vowed to keep exploring Pittsburgh by bicycle no matter how old I got, and planned an especially sweeping tour for my hundredth birthday in 2045. I had vowed to keep hating Amy in order to defy Mother, who kept prophesying I would someday not hate Amy. In short, I always vowed, one way or another, not to change. Not me. I needed the fierceness of vowing because I could scarcely help but notice, visiting the hatchling robins at school every day, that it was mighty unlikely.

As a life's work, I would remember everything—everything, against loss. I would go through life like a plankton net. I would trap and keep every teacher's funny remark, every face on the street, every microscopic alga's sway, every conversation, configuration of leaves, every dream, and every scrap of overhead cloud. Who would remember Molly's infancy if not me? (Unaccountably, I thought that only I had noticed—not Molly, but time itself. No one else, at least, seemed bugged by it. Children may believe that they alone have interior lives.)

Some days I felt an urgent responsibility to each change of light outside the sunporch windows. Who would remember any of it, any of this our time, and the wind thrashing the buckeye limbs outside? Somebody had to do it, somebody had to hang on to the days with teeth and fists, or the whole show had been in vain. That it was impossible never entered my reckoning. For work, for a task, I had never heard the word.

WE WERE MOVING THAT SPRING because our beloved grandfather had died, of a brain tumor. It was the year I got a microscope, and traveled with Judy Schoyer to Paw Paw, watched robins fledge from the school library window, and saw the Polyphemus moth walk toward Shadyside; I was now thirteen. I was expecting to attend an upper-school dance at the boys' school, Shady Side Academy, to which an older boy had invited me, but our grandfather died that day, and our father cried, and the dance was out. I shamed myself by minding that the dance was out.

Our grandfather had been a straitlaced, gentle man, whose mild and tolerant presence had soothed Oma for forty-four years. He doted on Amy and me, from a Scotch-Irish banker's distance; we loved him. For several weeks that spring as he lay in the hospital, the tumor pressed on his brain in such a way that he could say only one word, "Balls." Amy and I watched him move on the bed between sheets; he twisted inside a thin hospital gown. Neither of us had seen him angry before; he was angry now, and shocked. "Balls," he replied to any inquiry, "balls" for hello and "balls" for goodbye. Goodbye it was, and he died.

Oma sadly sold their Pittsburgh house. She and Mary Burinda moved into a pair of penthouse apartments in Shadyside. In the summer, Oma and Mary and Henry lived at Lake Erie. In the fall, they moved back to Pittsburgh and Oma caught opening nights at the Nixon Theater. And for the winter and spring, Oma and Mary moved to Pompano Beach, Florida, where they had an apartment on the water.

Oma sold their house, and we had bought it. A year later we moved from Richland Lane, from the generous house with the glass sunporch under buckeye trees. We moved to Oma's old stone house, another corner house, high on a hilly street where all the houses were old stone, and all their roofs were old slate, and the few children played—if they played at all—inside.

We lived now on a hushed hill packed with little castles. There were only three, dead-end streets. The longest street, winding silently into the very empyrean, was Glen Arden Drive. The McCulloughs lived at one end of it, and the McCradys at the other. Houses rarely changed hands; from here, there was nowhere in town to move to. The next step was a seat at the right hand of God.

In horizontal space, our family now lived near our first house on Edgerton Avenue, near St. Bede's Church. In vertical space, we were quite distant from it. Queerly, an inhumanly long and steep flight of outdoor stairs connected our newest neighborhood to our oldest one. These were the Glen Arden steps. From the top of Glen Arden Drive, between two houses, thirty concrete steps descended a scruffy unbuildable cliff to Dallas Avenue below, across the street from St. Bede's. The steps made a dark old tunnel. They were like the stairway in the poem, the stairway to the sea where down the blind are driven. Their concrete was so rough it ruined your shoes. Children from below played there; Popsicle wrappers and wrecked plastic toys kept the cliff a mess. People's maids used the steps to connect with Dallas Avenue buses—the maids climbed wearily up the cruel steps in the morning, and wearily down them at night.

The steps landed between and behind two small ordinary Pittsburgh brick houses; a walkway sloped down into the daylight of Dallas Avenue, where buses ran. If I stood on Dallas Avenue waiting for the bus to art class and looked across the street at the corner, by squinting down across the rows of sycamores at St. Bede's I could make out our first house. There it was on the farthest visible corner, painted white again, and there were the Lombardy poplars behind it, next to the alley. Down that now leafy street Jo Ann Sheehy had skated, and I had run from the nuns and run from the man whose windshield we hit with a snowball. There were the maple trees Henry planted when Amy and I were born.

Too old to play on the steps, instead I dreamed about them—how I dreamed about them!—a hundred steep steps dark as a chute. Fuzzy staghorn sumac poked their cold pipe rails. I dreamed about the blackened soil and frozen candy wrappers on the dizzying cliff they spanned. I dreamed the steps let me down in the wrong place. I dreamed the steps swayed underfoot, and rose, and tilted me over the ocean. I dreamed I couldn't find the steps.

That first spring our family walked out together, as we had not done for many years, to see the Memorial Day parade pass down Dallas Avenue. Our neighbors did, too; once a year the pale families emerged from their stone houses and climbed stiffly down the Glen Arden steps to watch the Memorial Day parade.

Now our family was seeing the once familiar parade from the other side of the street. A dozen bands passed, and the brass horns wagged from side to side in time. The kids from the big all-black high school came bopping and tossing batons. I felt the low drumbeats in my breastbone. More marching bands passed, then shuffling ranks of men, then children, in uniforms. Horses, of all things, walked by or skittered, backing the crowd. Then, to everyone's boredom, open cars drove by; it was over. Loose children on bikes, mad with excitement, rode squiggles and loops at the parade's rear, like tails on a kite. We stayed to hear the music wind away up toward the cemetery.

When the parade had passed, the people from the two sides of Dallas Avenue were left looking at each other. There were our former neighbors. Mother crossed over and talked to some of them. There were some of my earliest friends, altered, and my dear old friend Cathy Lindsey, whom I had already met up with again in our big public art classes; we always sat together. Now we waved across Dallas Avenue. The Sheehy women were there: Jo Ann in pink makeup, and her mother listless in a wide housedress, holding some sort of baby. The nice Fahey boys weren't there; they had moved. Father and I saw the polio boy who had worn such a tall shoe; now, miraculously, he had grown almost all the way up and had two equally long, good legs. He never knew us, so we didn't wave.

People were scattering. The Glen Arden families wordlessly climbed back up the thirty cement steps, and burst out like dead souls on another full scene enacted on a higher plane. They looked around strangely from between the two high houses, got their bearings so that the highest circle of Glen Arden Drive seemed like the very horizon, glided to their own houses, and closed their doors again.

SINCE WE HAD MOVED, my reading had taken a new turn.

Books wandered in and out of my hands, as they had always done, but now most of them had a common theme. This new theme was the source of imagination at its most private—never mentioned, rarely even brought to consciousness. It was, essentially, a time, and a series of places, to which I returned nightly. So also must thousands, or millions, of us who grew up in the 1950s, reading what came to hand. What came to hand in those years were books about the past war: the war in England, France, Belgium, Norway, Italy, Greece; the war in Africa; the war in the Pacific, in Guam, New Guinea, the Philippines; the war, Adolf Hitler, and the camps.

We read Leon Uris's popular novels, *Exodus,* and, better, *Mila 18,* about the Warsaw ghetto. We read Hersey's *The Wall*—again, the Warsaw ghetto. We read *Time* magazine, and *Life,* and *Look*. It was in the air, that there had been these things. We read, above all, and over and over, for we were young, Anne Frank's *The Diary of a Young Girl.* This was where we belonged; here we were at home.

I say "we," but in fact I did not know anyone else who read these things. Perhaps my parents did, for they brought the books home. What were my friends reading? We did not then talk about books; our reading was private, and constant, like the interior life itself. Still, I say, there must have been millions of us. The theaters of war—the lands, the multiple seas, the very corridors of air—and the death camps in Europe, with their lines of starved bald people . . . these, combined, were the settings in which our imaginations were first deeply stirred.

Earlier generations of children, European children, I inferred, had had on their minds heraldry and costumed adventure. They read *The Count of Monte Cristo* and *The Three Musketeers.* They read about King Arthur and Lancelot and Galahad; they read about Robin Hood. I had read some of these things and considered them behind me. It would have been pleasant, I suppose, to close your eyes and imagine yourself in a suit of armor, astride an armored horse, fighting a battle for honor with broadswords on a pennanted plain, or in a copse of trees.

But of what value was honor when, in book after book, the highest prize was a piece of bread? Of what use was a broadsword, or even a longbow, against Hitler's armies which occupied Europe, against Hitler's Luftwaffe, Hitler's Panzers, Hitler's U-boats, or against Hitler's S.S., who banged on the door and led Anne Frank and her family away? We closed our eyes and imagined how we would survive the death camps—maybe with honor and maybe not. We imagined how we would escape the death camps, imagined how we would liberate the death camps. How? We fancied and schemed, but we had read too much, and knew there was no possible way. This was a novel concept: Can't do. We were in for the duration. We closed our eyes and waited for the Allies, but the Allies were detained.

Now and over the next few years, the books appeared and we read them. We read *The Bridge Over the River Kwai, The Young Lions.* In the background sang a chorus of smarmy librarians:

> The world of books is a child's
> Land of enchantment.
> When you open a book and start reading
> You enter another world—the world
> Of make-believe—where anything can happen.

We read *Thirty Seconds Over Tokyo,* and *To Hell and Back.* We read *The Naked and the Dead, Run Silent, Run Deep,* and *Tales of the South Pacific,* in which American sailors saw native victims of elephantiasis pushing their own enlarged testicles before them in wheelbarrows. We read *The Caine Mutiny, Some Came Running.*

I was a skilled bombardier. I could run a submarine with one hand, and evade torpedoes, depth charges, and mines. I could disembowel a soldier with a bayonet, survive under a tarp in a lifeboat, and parachute behind enemy lines. I could contact the Resistance with my high-school French and eavesdrop on the Germans with my high-school German:

"Du! Kleines Mädchen! Bist du französisches Mädchen oder bist du Amerikanischer spy?"

"Je suis une jeune fille de la belle France, Herr S.S. Officer."

"Prove it!"

"Je suis, tu es, il est, nous sommes, vous êtes, ils sont."

"Very gut. Run along and play."

What were librarians reading these days? One librarian pressed on me a copy of *Look Homeward, Angel.* "How I envy you," she said, "having a chance to read this for the very first time." But it was too late, several years too late.

At last Hitler fell, and scientists working during the war came up with the atomic bomb. We read *On the Beach, A Canticle for Leibowitz;* we read *Hiroshima.* Reading about the bomb was a part of reading about the war: these were actual things and events, large in their effects on millions of people, vivid in their nearness to each man's or woman's death. It was a relief to turn from life to something important.

At school we had air-raid drills. We took the drills seriously; surely Pittsburgh, which had the nation's steel, coke, and aluminum, would be the enemy's first target.

I knew that during the war, our father, who was 4-F because of a collapsing lung, had "watched the skies." We all knew that people still watched the skies. But when the keen-eyed watcher spotted the enemy bomber over Pittsburgh, what, precisely, would be his moves? Surely he could only calculate, just as we in school did, what good it would do him to get under something.

When the air-raid siren sounded, our teachers stopped talking and led us to the school basement. There the gym teachers lined us up against the cement walls and steel lockers, and showed us how to lean in and fold our arms over our heads. Our small school ran from kindergarten through twelfth grade. We had air-raid drills in small batches, four or five grades together, because there was no room for us all against the walls. The teachers had to stand in the middle of the basement rooms: those bright Pittsburgh women who taught Latin, science, and art, and those educated, beautifully mannered European women who taught French, history, and German, who had landed in Pittsburgh at the end of their respective flights from Hitler, and who had baffled us by their common insistence on tidiness, above all, in our written work.

The teachers stood in the middle of the room, not talking to each other. We tucked against the walls and lockers: dozens of clean girls wearing green jumpers, green knee socks, and pink-soled white bucks. We folded our skinny arms over our heads, and raised to the enemy a clatter of gold scarab bracelets and gold bangle bracelets.

If the bomb actually came, should we not let the little kids—the kindergartners like Molly, and the first and second graders—go against the wall? We older ones would stand in the middle with the teachers. The European teachers were almost used to this sort of thing. We would help them keep spirits up; we would sing "Frère Jacques," or play Buzz.

Our house was stone. In the basement was a room furnished with a long wooden bar, tables and chairs, a leather couch, a refrigerator, a sink, an ice maker, a fireplace, a piano, a record player, and a set of drums. After the bomb, we would live, in the manner of Anne Frank and her family, in this basement. It had also a larger set of underground rooms, which held a washer and a dryer, a workbench, and, especially, food: shelves of canned fruits and vegetables, and a chest freezer. Our family could live in the basement for many years, until the radiation outside blew away. Amy and Molly would grow up there. I would teach them all I knew, and entertain them on the piano. Father would build a radiation barrier for the basement's sunken windows. He would teach me to play the drums. Mother would feed us and tend to us. We would grow close.

I had spent the equivalent of years of my life, I thought, in concentration camps, in ghettoes, in prison camps, and in lifeboats. I knew how to ration food and water. We would each have four ounces of food a day and eight ounces of water, or maybe only four ounces of water. I knew how to stretch my rations by hoarding food in my shirt, by chewing slowly, by sloshing water around in my mouth and wetting my tongue well before I swallowed. If the water gave out in the taps, we could drink club soda or tonic. We could live on the juice in canned food. I figured the five of us could live many years on the food in the basement—but I was not sure.

One day I asked Mother: How long could we last on the food in the basement? She did not know what I had been reading. How could she have known?

"The food in the basement? In the freezer and on the shelves? Oh, about a week and a half. Two weeks."

She knew, as I knew, that there were legs of lamb in the freezer, turkeys, chickens, pork roasts, shrimp, and steaks. There were pounds of frozen vegetables, quarts of ice cream, dozens of Popsicles. By her reckoning, that wasn't many family dinners: a leg of lamb one night, rice, and vegetables; steak the next night, potatoes, and vegetables.

"Two weeks! We could live much longer than two weeks!"

"There's really not very much food down there. About two weeks' worth."

I let it go. What did I know about feeding a family? On the other hand, I considered that if it came down to it, I would have to take charge.

It was clear that adults, including our parents, approved of children who read books, but it was not at all clear why this was so. Our reading was subversive, and we knew it. Did they think we read to improve our vocabularies? Did they want us to read and not pay the least bit of heed to what we read, as they wanted us to go to Sunday school and ignore what we heard?

I was now believing books more than I believed what I saw and heard. I was reading books about the actual, historical, moral world—in which somehow I felt I was not living.

The French and Indian War had been, for me, a purely literary event. Skilled men in books could survive it. Those who died, an arrow through the heart, thrilled me by their last words. This recent war's survivors, some still shaking, some still in mourning, taught in our classrooms. "*Wir waren ausgebommt,*" one dear old white-haired Polish lady related in German class, her family was "bombed out," and we laughed, we smart girls, because this was our slang for "drunk." Those who died in this war's books died whether they were skilled or not. Bombs fell on their cities or ships, or they starved in the camps or were gassed or shot, or they stepped on land mines and died surprised, trying to push their intestines back in their abdomens with their fingers and thumbs.

What I sought in books was imagination. It was depth, depth of thought and feeling; some sort of extreme of subject matter; some nearness to death; some call to courage. I myself was getting wild; I wanted wildness, originality, genius, rapture, hope. I wanted strength, not tea parties. What I sought in books was a world whose surfaces, whose people and events and days lived, actually matched the exaltation of the interior life. There you could live.

Those of us who read carried around with us like martyrs a secret knowledge, a secret joy, and a secret hope: There is a life worth living where history is still taking place; there are ideas worth dying for, and circumstances where courage is still prized. This life could be found and joined, like the Resistance. I kept this exhilarating faith alive in myself, concealed under my uniform shirt like an oblate's ribbon; I would not be parted from it.

We who had grown up in the Warsaw ghetto, who had seen all our families gassed in the death chambers, who had shipped before the mast, and hunted sperm whale in Antarctic seas; we who had marched from Moscow to Poland and lost our legs to the cold; we who knew by heart every snag and sandbar on the Mississippi River south of Cairo, and knew by heart Morse code, forty parables and psalms, and lots of Shakespeare; we who had battled Hitler and Hirohito in the North Atlantic, in North Africa, in New Guinea and Burma and Guam, in the air over London, in the Greek and Italian hills; we who had learned to man minesweepers before we learned to walk in high heels—were we going to marry Holden Caulfield's roommate, and buy a house in Point Breeze, and send our children to dancing school?

THE BOYS WERE CHANGING. Those froggy little beasts had elongated and transformed into princes and gods. When it happened, I must have been out of the room. Suddenly here they all were, Richie and Rickie and Dan and all, diverse in their varied splendors, each powerful and mysterious, immense, and possessed of an inexplicable knowledge of arcana.

The boys wandered the neighborhoods now, and showed up at girls' houses, as if by accident. They would let us listen to them talk, and we heard them mention the state legislature, say, or some opinion of Cicero's, or the Battle of the Marne—and those things abruptly became possible topics in society because those magnificent boys had pronounced their names.

Where had they learned all this, or, more pertinently, why had they remembered it? We girls knew precisely the limits of the possible and the thinkable, we thought, and were permanently astonished to learn that we were wrong. Whose idea of sophistication was it, after all, to pay attention in Latin class? It was the boys' idea. Everything was. Everything they thought of was bold and original like that. While we were worried about sending valentines, they were worried about sending troops.

Plus their feet were so big. You could look at the boys' sheer physical volume with uncomprehending astonishment forever. Had the braces on their teeth been restraining their very bones? For look at them. You would never tire of running your wondering eyes over the mystery of their construction, so plain, and the mystery of their bulk, and the mystery of their skin, and even their strange boxy clothes. The boys.

We ran, we fancied, to sweetness, we girls. The boys, as we got to know them, were cynical. They addressed each other out of the corners of their mouths in cryptic staccato phrases, all clever references to that larger world wherein they dwelt and where we longed to go ourselves. If you got to know them, apparently, they would tell you about their teachers at Shady Side Academy—teachers my own father had studied under, but about whom, alas, he could come up with precious little.

We girls chafed, whined, and complained under our parents' strictures. The boys waged open war on their parents. They cursed their fathers, and disobeyed them outright. ("What can they do? Throw me out?") Was this not breathtakingly bold? The boys' pitched battles with their parents were legendary; the punishments they endured melted our hearts.

Each year as we rose through the grades, dancing school met an hour later, until one year it vanished into the darkness, and was replaced by, or transmogrified into, another institution altogether, that of country-club subscription dances.

The engraved invitation came in the mail: The Sewickley Country Club was hosting a subscription dinner dance several weeks thence. Each of several appropriate country clubs, it turned out, gave precisely one such dance a year, at a time that coincided with boarding-school vacations. I knew Sewickley children, having opposed on playing fields their school's ferocious field hockey team. The old village of Sewickley had come to prominence late in the nineteenth century when some families quit their grandparents' mansions on Fifth Avenue and moved in a body to that green and pleasant land. They zoned it to a fare-thee-well and furnished it with a country club, a Presbyterian church, and a little expensive school. Now they were asking us to a dinner dance.

We showed up at our own country club in pale spaghetti-strap dresses and silk shoes, to board a yellow bus in the snow. There we all were: the same boys, the same girls. How did they know? I wondered which of those remote country-club powers, those white-haired sincere men, those golden-haired, long-toothed, ironic women, had met on what firm cloud over western Pennsylvania to apportion and schedule these events among the scattered country clubs, and had pored over what unthinkable list of schoolchildren to discuss which schoolchildren should be asked to these dances they held for what reason. If you were part Jewish, would they find you out, like Hitler? How small a part could they detect? What was at the end of all this novitiate—solemn vows?

We dined that night in faraway Sewickley, at long linen-covered tables marked by place cards. Our shrimp cocktails were already at our places. We were like Beauty in the castle of the Beast: that is, I, at least, never laid eyes on the unknown adult or adults who had presumably invited us, designed and ordered the invitations, secured a room and a band, and devised the menu. There were some adults against the walls, all dressed up, who ignored us and whom we ignored.

My dinner partner was a fragile redhead from Sewickley, a Paulie—from St. Paul's School—whose hulking twin sisters had several times mown me down on the hockey field. From him I learned that some girls my age voluntarily played golf. Like many of the boys, he was good-natured, polite, somewhat cowed, and delicately handsome. One was not, however, thank God, required to fall in love at a subscription dance, although it had been known to happen.

We ate chicken breast in velvet sauce on ham. We ate wild rice, tomato aspic, and, as a concession to our being in fact children, hot fudge sundaes or green peppermint parfaits.

During dessert the band straggled in and set up by the freezing French doors to the terrace. The band was an unmatched set of bored men in dark suits and red carnations. The only bands that counted in our book were Lester Lanin's and Eddy Duchin's. These men, as at all subscription dances, were merely locals: a drum, a bass, a piano, a clarinet. Their boredom, and the possible death of their musical ambitions, and the probable complete disregard of everyone with whom they had dealt over this engagement, unless they had had the good fortune to run into my mother, had drained all the expression from their faces. Sometimes, though, on a jitterbug or a Charleston, you could pry a wink out of the drummer.

The band struck up, not surprisingly, "Mountain Greenery." This frenzied sequence of notes had been our cue conditioned since we were ten. We danced.

There were boys here from far away—not only from familiar Fox Chapel and expected Sewickley, but also from Ligonier, that pretty village in the distant mountains where the Mellons lived. There were older boys here, who had already been to deb parties. And there were some very tall boys—some of ours and some of theirs—whose shoulders rose above our heads like those few lone trees which burst through the canopy in a rain forest. Although these big boys' status was as great as their stature, they rarely smiled or relaxed, but instead looked worriedly around over our headtops, frowning, earnest, always at the edge of a wince.

"Isn't he cute?"

In the densely carpeted ladies' rooms we all hurried. We didn't meet each other's eyes in the long mirrors.

"Which?"

"Which what?"

"Which is cute?"

It was always one of the wincing giants who was cute. We ran combs through our hair and pounded back along the labyrinthine club corridors to the dance floor. Yes, very cute.

One blond, sharp-toothed boarding-school boy, a famously witty chess player, was wearing patent-leather pumps. On his feet, that is, where his shoes should be, he was wearing low-slung, dainty, shiny pumps, like ballet practice shoes, with satin bows at the toes—and he carried it off. Thus I learned yet again that more things were possible in the world than I had dreamed. He and a friend had driven a car through the snow to this dance. The friend was a sarcastic boy, narrow-skulled and overbred as a collie, who said he hung around in the Hill District. The Hill District was Pittsburgh's cruelest and coolest black ghetto, where more babies died than anywhere else in the United States. Up on the Hill, he went to whorehouses. Was this not bold, evil, original? Our own boys would never think of that.

I had sat near these two at dinner. They had traveled. From their boarding school they had walked, loose, in the towns of Connecticut, and knew them well enough to dismiss them. I danced with each of them. How light the blond boy's shoulders felt! With what smooth disdain did the blond boy lead me walking beside him four steps before he pulled me in again to him, as easily as if my arm had been the bowline of a boat!

And we were buoyant when we danced, we two, were we not? Had he noticed?

This light-shouldered boy could jitterbug, old style, and would; he was more precious than gold, yea, than much fine gold. We jitterbugged. There was nothing flirtatious about it. It was more an exultant and concentrated collaboration, such as aerialists enjoy—and I hope they enjoy it—when they catch each other twirling in midair. Only the strength in our fingertips kept us alive. If they weakened or slipped, his fingertips or mine, we'd fall spinning backward across the length of the room and out through the glass French doors to the snowy terrace, and if we were any good we'd make sure we fell on the downbeat, snow or no snow. For this was, at last, rock and roll. We danced in front of the band; I wished the music were louder.

The last dance was slow; the lights dimmed. The light-shouldered blond boy moved me over and across the golden dark floor and in and out of his arms. He released me and caught me, slowly, and turned me and spun me, and paused on the odd long note so I had to raise a leg from the hip to keep us afloat, and I held him loosely but surely for the count of four, amazed.

He was bare-handed, as were all the boys at these dances. We retained our white cotton gloves. It was easier now to imagine his warmth, the heel of his naked left hand on my glove. But it still required imagination. The thick cotton stretched flat across the dip of my palm like a trampoline; it repelled the bulge of his hand and held away his heat.

"Keep your back straight," my mother had told me years ago. "Don't let your arm weigh and drag on a boy's shoulder, no matter how tired you are. Dance on the balls of your feet, no matter how tall you are. Chin up."

The drummer stretched in the dark and rubbed the back of his neck. He began packing up, retaining, however, his brushes for "Good Night, Ladies," at whose opening bars we all groaned.

We groaned because we had to part and lacked the words to manage it smoothly. We groaned because we had to ride back through the snow for an hour and a half with our boys on a bus, and we never figured out how to conduct ourselves on this bus. Were we to kiss, or sing camp songs?

"How was it?" my mother asked the next morning. She lowered the Sunday paper she'd been paging through. How was what? I could barely remember. Someone's father had picked us up at our club and driven us another hour home. I didn't get in till after two. Now it was Sunday morning. I was dressed up again and looking for a pair of clean white cotton gloves for church. So was Amy. If there was such a pair, I wanted to find it first.

"How was it?" she asked, and then I remembered and began to understand how it was. It was wonderful, that's how it was. It was absolutely wonderful.

THAT MORNING IN CHURCH after our first subscription dance, we reconvened on the balcony of the Shadyside Presbyterian Church. I sat in the first balcony row, and resisted the impulse to stretch my Charleston-stiff legs on the balcony's carved walnut rail. The blond boy I'd met at the dance was on my mind, and I intended to spend the church hour recalling his every word and gesture, but I couldn't concentrate. Beside me sat my friend Linda. Last night at the dance she had been a laughing, dimpled girl with an advanced sense of the absurd. Now in church she was grave, and didn't acknowledge my remarks.

Near us in the balcony's first row, and behind us, were the boys—the same boys with whom we had traveled on a bus to and from the Sewickley Country Club dance. Below us spread the main pews, filling with adults. Almost everyone in the church was long familiar to me. But this particular Sunday in church bore home to me with force a new notion: that I did not really know any of these people at all. I thought I did—but, being now a teenager, I thought I knew almost everything. Only the strongest evidence could penetrate this illusion, which distorted everything I saw. I knew I approved almost nothing. That is, I liked, I adored, I longed for, everyone on earth, especially India and Africa, and particularly everyone on the streets of Pittsburgh—all those friendly, democratic, openhearted, sensible people—and at Forbes Field, and in all the office buildings, parks, streetcars, churches, and stores, excepting only the people I knew, none of whom was up to snuff.

The church building, where the old Scotch-Irish families

assembled weekly, was a Romanesque chunk of rough, carved stone and panes of dark slate. Covered in creeper, long since encrusted into its quietly splendid site, it looked like a Scottish rock in the rain.

Everywhere outside and inside the church and parish hall, sharp carved things rose from the many dim tons of stone. There were grainy crossed keys, pelicans, anchors, a phoenix, ivy vines, sheaves of wheat, queer and leering mammal heads like gargoyles, thistles for Scotland, lizards, scrolls, lions, and shells. It looked as if someone had once in Pittsburgh enjoyed a flight or two of fancy. If your bare hand or arm brushed against one of the stone walls carelessly, the stone would draw blood.

My wool coat sat empty behind me; its satin lining felt cool on the backs of my arms. I hated being here. It looked as if the boys did, too. Their mouths were all open, and their eyelids half down. We were all trapped. At home before church, I had been too rushed to fight about it.

I imagined the holy war each boy had fought with his family this morning, and lost, resulting in his sullen and suited presence in church. I thought of Dan there, ruddy-cheeked, and of wild, sweet Jamie beside him, each flinging his silk tie at his hypocrite father after breakfast, and making a desperate stand in some dark dining room lighted upward by snowlight from the lawns outside—struggling foredoomed to raise the stone and walnut weight of this dead society's dead institutions, battling for liberty, freedom of conscience, and so forth.

The boys, at any rate, slumped. Possibly they were hung over.

While the nave filled we examined, or glared at, the one thing before our eyes: the apse's enormous gold mosaic of Christ. It loomed over the chancel; every pew in the nave and on the balcony looked up at it. It was hard to imagine what long-ago board of trustees had voted for this Romish-looking mosaic, so glittering, with which we had been familiarizing ourselves in a lonely way since infancy, when our eyes could first focus on distance.

Christ stood barefoot, alone and helpless-looking, his

palms outcurved at his sides. He was wearing his robes. He wasn't standing on anything, but instead floated loose and upright inside a curved, tiled dome. The balcony's perspective foreshortened the dome's curve, so Christ appeared to drift flattened and clumsy, shriveled but glorious. Barefoot as he was, and with the suggestion of sandstone scarps behind him, he looked rural. Below me along the carpeted marble aisles crept the church's families; the women wore mink and sable stoles. Hushed, they sat and tilted their hatted heads and looked at the rural man. His skies of shattered gold widened over the sanctuary and almost met the square lantern tower, gold-decorated, over the nave.

The mosaic caught the few church lights—lights like tapers in a castle—and spread them dimly, a dusting of gold like pollen, throughout the vast and solemn space. There was nothing you could see well in this rich, Rembrandt darkness—nothing save the minister's shining face and Christ's gold vault—and yet there was no corner, no scratchy lily work, you couldn't see at all.

It was a velvet cord, maroon, with brass fittings, that reserved our ninth-grade balcony section for us. We sat on velvet cushions. Below us, filling the yellow pews with dark furs, were the rest of the families of the church, who seemed to have been planted here in dignity—by a God who could see how hard they worked and how few pleasures they took for themselves—just after the Flood went down. There were Linda's parents and grandparents and one of her great-grandparents. Always, the same old Pittsburgh families ran this church. The men, for whose forefathers streets all over town were named, served as deacons, trustees, and elders. The women served in many ways, and ran the Christmas bazaar.

I knew these men; they were friends and neighbors. I knew what they lived for, I thought. The men wanted to do the right thing, at work and in the community. They wore narrow, tight neckties. Close-mouthed, they met, in volunteer boardrooms and in club locker rooms, the same few comfortable others they had known since kindergarten. Their wives and children, in those days, lived around them on their visits home. Some men found their families bewildering, probably; a man might wonder, wakened by reports of the outstanding misdeeds of this son or that son, how everyone had so failed to understand what he expected. Some of these men held their shoulders and knuckles tight; their laughter was high and embarrassed; they seemed to be looking around for the entrance to some other life. Only some of the doctors, it seemed to me, were conspicuously interested and glad. During conversations, they looked at people calmly, even at their friends' little daughters; their laughter was deep, long, and joyful; they asked questions; and they knew lots of words.

I knew the women better. The women were wise and strong. Even among themselves, they prized gaiety and irony, gaiety and irony come what may. They coped. They sighed, they permitted themselves a remark or two, they lived essentially alone. They reared their children with their own two hands, and did all their own cooking and driving. They had no taste for waste or idleness. They volunteered their considerable energies, wisdom, and ideas at the church or the hospital or the service organization or charity.

Life among these families partook of all the genuine seriousness of life in time. A child's birth was his sole entrée, just as it is to life itself. His birthright was a regiment of families and a phalanx of institutions which would accompany him, solidly but at a distance, through this vale of tears.

Families whose members have been acquainted with each other for as long as anyone remembers grow not close, but respectful. They accumulate dignity by being seen at church every Sunday for the duration of life, despite their troubles and sorrows. They accumulate dignity at club luncheons, dinners, and dances, by gracefully and persistently, with tidy hair and fitted clothes, occupying their slots.

In this world, some grown women went carefully wild from time to time. They appeared at parties in outlandish clothes, hair sticking out, faces painted in freckles. They shrieked, sang, danced, and parodied anything—that is, anything at all outside the tribe—so that nothing, almost, was sacred. These clowns were the best-loved women, and rightly so, for their own sufferings had taught them what dignity was worth, and every few years they reminded the others, and made them laugh till they cried.

My parents didn't go to church. I practically admired them for it. Father would drive by at noon and scoop up Amy and me, saying, "Hop in quick!" so no one would see his weekend khaki pants and loafers.

Now, in unison with the adults in the dimness below, we read responsively, answering the minister. Our voices blended low, so their joined sound rose muffled and roaring, rhythmic, like distant seas, and soaked into the rough stone vaults and plush fittings, and vanished, and rose again:

> The heavens declare the glory of God:
> AND THE FIRMAMENT SHOWETH HIS HANDYWORK.
> Day unto day uttereth speech,
> AND NIGHT UNTO NIGHT SHOWETH KNOWLEDGE.
> There is no speech nor language, where their voice is not heard.

The minister was a florid, dramatic man who commanded a batch of British vowels, for which I blamed him absolutely, not knowing he came from a Canadian farm. His famous radio ministry attracted letters and even contributions from Alaskan lumberjacks and fishermen. The poor saps. What if one of them, a lumberjack, showed up in Pittsburgh wearing a lumberjack shirt and actually tried to enter the church building? Maybe the ushers were really bouncers.

I had got religion at summer camp, and had prayed nightly there and in my bed at home, to God, asking for a grateful heart, and receiving one insofar as I requested it. Inasmuch as I despised everything and everyone about me, of course, it was taken away, and I was left with the blackened heart I had chosen instead. As the years wore on, the intervals between Julys at camp stretched, and filled with country-club evenings, filled with the slang of us girls, our gossip, and our intricately shifting friendships, filled with the sight of the boys whose names themselves were a litany, and with the absorbing study of their nonchalance and gruff ease. All of which I professed, from time to time, when things went poorly, to disdain.

Nothing so inevitably blackened my heart as an obligatory Sunday at the Shadyside Presbyterian Church: the sight of orphan-girl Liz's "Jesus" tricked out in gilt; the minister's Britishy accent; the putative hypocrisy of my parents, who forced me to go, though they did not; the putative hypocrisy of the expensive men and women who did go. I knew enough of the Bible to damn these people to hell, citing chapter and verse. My house shall be called the house of prayer; but ye have made it a den of thieves. Every week I had been getting madder; now I was going to plain quit. One of these days, when I figured out how.

After the responsive reading there was a pause, an expectant hush. It was the first Sunday of the month, I remembered, shocked. Today was Communion. I would have to sit through Communion, with its two species, embarrassment and tedium—and I would be late getting out and Father would have to drive around the block a hundred times. I had successfully avoided Communion for years.

From their pews below rose the ushers and elders—everybody's father and grandfather, from Mellon Bank & Trust et cetera—in tailcoats. They worked the crowd smoothly, as always. When they collected money, I noted, they were especially serene. Collecting money was, after all, what they did during the week; they were used to it. Down each pew an usher thrust a long-handled velvet butterfly net, into the invisible interior of which we each inserted a bare hand to release a crushed, warm dollar bill we'd stored in a white glove's palm.

Now with dignity the ushers and elders hoisted the round sterling silver trays which bore Communion. A loaded juice tray must have weighed ten pounds. From a cunning array of holes in its top layer hung wee, tapered, lead-crystal glasses. Each held one-half ounce of Welch's grape juice.

The seated people would pass the grape-juice trays down the pews. After the grape juice came bread: flat silver salvers bore heaps of soft bread cubes, as if for stuffing a turkey.

The elders and ushers spread swiftly and silently over the marble aisles in discreet pairs, some for bread cubes, some for grape juice, communicating by eyebrow only. An unseen organist, behind stone screens, played a muted series of single notes, a restless, breathy strain in a minor key, to kill time.

Soon the ushers reached the balcony where we sat. There our prayers had reached their intensest pitch, so fervent were we in our hopes not to drop the grape-juice tray.

I passed up the Welch's grape juice, I passed up the cubed bread, and sat back against my coat. Was all this not absurd? I glanced at Linda beside me. Apparently it was not. Her hands lay folded in her lap. Both her father and her uncle were elders.

It was not surprising, really, that I alone in this church knew what the barefoot Christ, if there had been such a person, would think about things—grape juice, tailcoats, British vowels, sable stoles. It was not surprising because it was becoming quite usual. After all, I was the intelligentsia around these parts, single-handedly. The intelligentsium. I knew why these people were in church: to display to each other their clothes. These were sophisticated men and women, such as we children were becoming. In church they made business connections; they saw and were seen. The boys, who, like me, were starting to come out for freedom and truth, must be having fits, now that the charade of Communion was in full swing.

I stole a glance at the boys, then looked at them outright, for I had been wrong. The boys, if mine eyes did not deceive me, were praying. Why? The intelligentsia, of course, described itself these days as "agnostic"—a most useful word. Around me, in seeming earnest, the boys prayed their unthinkable private prayers. To whom? It was wrong to watch, but I watched.

On the balcony's first row, to my right, big Dan had pressed his ruddy cheeks into his palms. Beside him, Jamie bent over his knees. Over one eye he had jammed a fist; his other eye was crinkled shut. Another boy, blond Robert, lay stretched over his arms, which clasped the balcony rail. His shoulders were tight; the back of his jacket rose and fell heavily with his breathing. It had been a long time since I'd been to Communion. When had this praying developed?

Dan lowered his hands and leaned back slowly. He opened his eyes, unfocused to the high, empty air before him. Wild Jamie moved his arm; he picked up a fistful of hair from his forehead and held it. His eyes fretted tightly shut; his jaws worked. Robert's head still lay low on his outstretched sleeves; it moved once from side to side and back again. So they struggled on. I finally looked away.

Below the balcony, in the crowded nave, men and women were also concentrating, it seemed. Were they perhaps pretending to pray? All heads were bent; no one moved. I began to doubt my own omniscience. If I bowed my head, too, and shut my eyes, would this be apostasy? No, I'd keep watching the people, in case I'd missed some clue that they were actually doing something else—bidding bridge hands.

For I knew these people, didn't I? I knew their world, which was, in some sense, my world, too, since I could not, outside of books, name another. I knew what they loved: their families, their houses, their country clubs, hard work, the people they knew best, and summer parties with old friends full of laughter. I knew what they hated: labor unions, laziness, spending, wildness, loudness. They didn't buy God. They didn't buy anything if they could help it. And they didn't work on spec.

Nevertheless, a young father below me propped his bowed head on two fists stacked on a raised knee. The ushers and their trays had vanished. The people had taken Communion. No one moved. The organist hushed. All the men's heads were bent—black, white, red, yellow, and brown. The men sat absolutely still. Almost all the women's heads were bent down, too, and some few tilted back. Some hats wagged faintly from side to side. All the people seemed scarcely to breathe.

I was alert enough now to feel, despite myself, some faint, thin stream of spirit braiding forward from the pews. Its flawed and fragile rivulets pooled far beyond me at the altar. I felt, or saw, its frail strands rise to the wide tower ceiling, and mass in the gold mosaic's dome.

The gold tesserae scattered some spirit like light back over the cavernous room, and held some of it, like light, in its deep curve. Christ drifted among floating sandstone ledges and deep, absorbent skies. There was no speech nor language. The people had been praying, praying to God, just as they seemed to be praying. That was the fact. I didn't know what to make of it.

I left Pittsburgh before I had a grain of sense. Who IS my neighbor? I never learned what the strangers around me had known and felt in their lives—those lithe, sarcastic boys in the balcony, those expensive men and women in the pews below—but it was more than I knew, after all.

YEARS BEFORE THIS, on long-ago summer Sundays, before Father went down the Ohio and ended up selling his boat, he used to take me out with him on the water. It was a long drive to the Allegheny River; it was a long wait, collecting insects in the grass among the pebbles on shore, till Father got the old twenty-four-foot cabin cruiser ready to go. But the Allegheny River, once we got out on it, was grand. Its distant shores were mostly wooded on both sides; coal barges, sand barges, and shallow-draft oil tankers floated tied up at a scattering of docks. Father wore tennis shoes on his long feet, and a sun-bleached cotton captain-style hat. He always squinted outside, hat or no hat, because his eyes were such a pale blue; the sun got in them. He was so tall he had to lean under the housetop to man the wheel.

We stopped at islands and swam. There were wooded islands in the river—like Smoky Island at Pittsburgh's point, where Indians had tortured their English and Scotch-Irish captives by night. The Indians had tied the soldiers and settlers to trees, heaped hot coals on their feet, and let their small boys practice archery on them. Indian women heated rifle barrels and ramrods over fires till they glowed, then drove them through prisoners' nostrils or ears. The screams of the tortured settlers on Smoky Island reached French soldiers at Fort Duquesne, who had handed them over to the Indians reluctantly, they said. "Humanity groans at being forced to use such monsters."

Father and I tied up at Nine-Mile Island, upstream from Smoky Island, and I jumped from a high rope-swing into the water, after poor Father told me all about those boaters' children who'd been killed or maimed dropping from this very swing. He could not bear to watch; he shut his eyes. From the tree branch at the top of the ladder I jumped onto the swing; when I let go over the water, momentum shot me forward like a slung stone. I swam up to find the water's surface again, and called to Father onshore, "It's okay now."

Our boat carved through the glossy water. Pittsburgh's summer skies are pale, as they are in many river valleys. The blinding haze spread overhead and glittered up from the river. It was the biggest sky in town.

We rode up in the locks and down in the locks. The locks scared me, for the huge doors that locked out the river leaked, and loud tons of water squirted in, and we sat helpless below the river with nothing to do but wait for the doors to give way. Enormous whirlpools dragged at the boat; we held on to the lock walls, clawed, with a single hand line and a boat hook. Once I dropped the boat hook, a new one with a teak handle, and the whirlpools sucked it down. To where? Where did the whirlpools put the water they took, and where would they put you, all ground up, if you fell in?

Oh, the river was grand. Outside the lock and back on the go, I sang wild songs at the top of my voice out over the roaring boat's stern. We raced under old steel bridges set on stone pilings in the river. How do people build bridges? How did anyone set those pilings, pile those stones, under the water?

Whenever I was on the river, I seemed to be visiting a fascinating place I had forgotten all about, where physical causes had physical effects, and great things got done, slowly, heavily, because people understood materials and forces.

Father on these boat outings answered my questions at length. He explained that people built coffer dams to set bridge pilings in a river. They lowered a kind of big pipe, or tight set of walls, to the bottom, and pumped all the water out of it; then the men could work there. I imagined the men piling and mortaring stones, with the unhurried ease of stone masons; they stood on gasping catfish and stinky silt. They were working under the river, at the bottom of a well of air. Just a few inches away, outside their coffer dam, a complete river of water was sliding downhill from western New York to the Gulf of Mexico. Above the workers' heads, boats and barges went by, their engines probably buzzing the coffer-dam walls. What a life. Father said that some drowned in accidents, or got crushed; it was dangerous work. He said, answering my question, that these workers made less money than the men I knew, men I privately considered wholly unskilled. The bridge pilings obsessed me; I thought and thought about the brave men who built them in the rivers. I tried to imagine their families, their lunches, their boots. I tried to imagine what it would feel like to accomplish something so useful as building a bridge. What a queer world was the river, where I admired everything and knew nothing.

Father explained how to make glass from sand. He explained, over and over, because I was usually too frightened to hear right, how the river locks worked; they ran our boat up or down beside the terrible dams. The concrete navigation dams made slick spillways like waterfalls across the river. From upstream it was hard to see the drop's smooth line. Drunks forgot about the dams from time to time, and drove their boats straight over, killing themselves and everyone else on board. How did the drunks feel, while they were up loose in the air at the wheels of their boats for a split second, when they remembered all of a sudden the dam? "Oh yes, the dam." It seemed like a familiar feeling.

On the back of a chart—a real nautical chart, with shoals and soundings, just as in *Life on the Mississippi*—Father drew a diagram of a water system. The diagram made clear something I'd always wondered about: how water got up to the top floors of houses. The water tower was higher than the highest sinks, that was all; through all those labyrinthine pipes, the water sought its own level, seeming to climb up, but really still trickling down. He explained how steam engines worked, and suspension bridges, and pumps.

Father explained so much technology to me that for a long time I confused it with American culture. If pressed, I would have claimed that an American invented the irrigation ditch. Certainly the coffer dam was American, I thought, and the water tower, the highway tunnel—these engineering feats—and everything motorized, and everything electrical, and in short, everything I saw about me newer than fishnets, sailboats, and spoons.

Technology depended on waterworks. The land of the forty-eight states was an extended and mighty system of controlled slopes, a combination Grand Coulee Dam and Niagara Falls. The water fell and the turbines spun and the lights came on, so steel mills could run all night. Then the steel made cars, millions of cars, and workers bought the cars, because Henry Ford in 1910 had come up with the idea of paying them enough to buy things. So the water rolled down the continent—just plain fell—and everyone got rich.

Now, years later, Father had picked Amy and me up after church. When we got out of the car in the garage, we could hear Dixieland, all rambling brasses and drums, coming from the house. We hightailed it inside through the snow on the back walk and kicked off our icy dress shoes. I was in stockings. I could eat something, and go to my room. I had my own room now, and when I was home I stayed there and read or sulked.

While we were making sandwiches, though, Father started explaining the world to us once again. I stuck around. There in the kitchen, Father embarked upon an explanation of American economics. I don't know what prompted it. His voice took on urgency; he paced. Money worked like water, he said.

We were all listening, even little Molly. Molly, at four, had an open expression, smooth and quick, and fine blond hair; she was eating on the hoof, like the rest of us, and looking up, a pale face at thigh level, following the conversation. Mother futzed around the kitchen in camel-colored wool slacks; she rarely ate.

Did we know how water got up to our attic bathroom? Money worked the same way, he said, worked the way locks on the river worked, worked the way water flowed down from high water towers into our attic bathroom, the way the Allegheny and the Monongahela flowed into the Ohio, and the Ohio flowed into the Mississippi and out into the Gulf of Mexico at New Orleans. The money, once you got enough of it high enough, would flow by gravitation, all over everybody.

"It doesn't work that way," our mother said. She offered Molly tidbits: a drumstick, a beet slice, cheese. "Remember those shacks we see in Georgia? Those barefoot little children who have to quit school to work in the fields, their poor mothers not able to feed them enough"—we could all hear in her voice that she was beginning to cry—"not even able to keep them dressed?" Molly was looking at her, wide-eyed; she was bent over looking at Molly, wide-eyed.

"They shouldn't have so many kids," Father said. "They must be crazy."

The trouble was, I no longer believed him. It was beginning to strike me that Father, who knew the real world so well, got some of it wrong. Not much; just some.

Part Three

PITTSBURGH WASN'T REALLY ANDREW CARNEGIE'S TOWN. We just thought it was. Steel wasn't the only major industry in Pittsburgh. We just had to think to recall the others.

Andrew Carnegie started out in Pittsburgh as a tiny bobbin boy, and ended up a tiny millionaire; he was only five feet three. When he was twenty-four, having scrambled, he became superintendent of the Western Division of the Pennsylvania Railroad. Whenever wrecks blocked the railroad tracks, Carnegie showed up to supervise. He hopped around the wrecked freight cars; he ordered the big workmen to lay tracks around the wrecks or even, quick, to burn the wrecks to save the schedule. He liked to tell about one such night, when an enormous, unknowing Irish workman picked him straight up off the ground and set him aside like a gate, booming at him, "Get out of the way, you brat of a boy. You're eternally in the way of the men who are trying to do their job."

The Carnegies emigrated from Scotland when Andrew was thirteen. A bookish family of Lowland Scots radicals, they championed universal suffrage, and hated privilege and hereditary wealth. "As a child," he recalled, "I could have slain king, duke, or lord, and considered their death a service to the state." When later Edward VII offered him a title, he refused it.

The then fashionable suburb of Homewood, where young Carnegie moved with his mother in 1859, was part of an old estate. The center of life there was the estate house of eighty-year-old Judge William Wilkins and his wife, Mathilda. Wilkins had served in government under three Presidents and returned to Pittsburgh; Mathilda Wilkins was from a prominent family whose members had served in two cabinets. The Civil War was then heating up, and the talk one social evening was of Negroes. Young Carnegie was among the guests. Mrs. Wilkins complained of Negroes' "forwardness." It was disgraceful, she said: Negroes admitted to West Point.

"Oh, Mrs. Wilkins," Carnegie piped up. He was then only in his twenties, but a man of convictions, which he didn't shed when he visited the great house. "There is something even worse than that. I understand that some of them have been admitted to heaven!"

"There was a silence that could be felt," Carnegie recalled. "Then dear Mrs. Wilkins said gravely:

" 'That is a different matter, Mr. Carnegie.' "

Carnegie started making steel. He wrote four books. He preached what he called, American style, the Gospel of Wealth. A man of wealth should give it away for the public good, and not weaken his sons with it. "The man who dies rich, dies disgraced."

In 1901, when he was sixty-six, Carnegie sold the Carnegie Company to J. P. Morgan, for $480 million. His share came to $250 million. Carnegie added this sum to his considerable other wealth—he had to build a special steel room in Hoboken, New Jersey, to house the bulky paper bonds, pesky things—and set about giving it away. He managed to get rid of $350 million of it before he died, in 1919, leaving for himself while he lived, and his family when he died, very much less than a tithe.

Carnegie's top steelmen were share-owning partners—forty of them—most of whom had worked their way up from the blast furnaces, smelters, and rolling mills. When J. P. Morgan bought the company he called U.S. Steel, these forty split the rest of the take, and became instant millionaires. One went to a barber on Penn Avenue for his first shampoo; the barber reported that the washing "brought out two ounces of fine Mesabi ore and a scattering of slag and cinders."

Carnegie gave over $40 million to build 2,509 libraries.

All the early libraries had graven over their doors: LET THERE BE LIGHT.

But a steelworker, speaking for many, told an interviewer, "We didn't want him to build a library for us, we would rather have had higher wages." At that time steelworkers worked twelve-hour shifts on floors so hot they had to nail wooden platforms under their shoes. Every two weeks they toiled an inhuman twenty-four-hour shift, and then they got their sole day off. The best housing they could afford was crowded and filthy. Most died in their forties or earlier, from accidents or disease. Workers' lives were almost unbearable in Düsseldorf then, too, and in Lisle, and Birmingham, and Ghent. It was the Gilded Age.

While Carnegie was visiting Scotland in 1892, his man Henry Clay Frick had loosed three hundred hired guns—Pinkertons—on unarmed strikers and their families at the Homestead plant up the river, strikers who subsequently beat the daylights out of Pinkertons with their fists. Frick then called in the entire state militia, eight thousand strong, whose armed occupation of the Homestead plant not only broke the strike but also killed all unions in the steel industry nationwide until 1936.

Pittsburgh's astounding wealth came from iron and steel, and also from aluminum, glass, coke, electricity, copper, natural gas—and the banking and transportation industries that put up the money and moved the goods. Some of the oldest Scotch-Irish and German families in Pittsburgh did well, too, like the sons of Scotch-Irish Judge Mellon. Andrew Mellon, a banker, invested in aluminum when the industry consisted of a twenty-two-year-old Oberlin College graduate who made it in his family's woodshed. He also invested in coke, iron, steel, and oil. When he was named Secretary of the Treasury, quiet Andrew Mellon was one of three Americans who had ever amassed a billion dollars. (Carnegie's strategy was different; he followed the immortal dictum: "Put all your eggs in the one basket and—*watch that basket.*")

By the turn of the century, Pittsburgh had the highest death rate in the United States. That was the year before Carnegie sold his steel company. Typhoid fever epidemics recurred, because Pittsburgh's council members wouldn't filter the drinking water; they disliked public spending. Besides, a water system would mean a dam, and a dam would yield cheap hydroelectric power, so the power companies would buy less coal; coal-company owners and their bankers didn't want any dams. Pittsburgh epidemics were so bad that boatmen on the Ohio River wouldn't handle Pittsburgh money, for fear of contagion.

While Carnegie was unburdening himself publicly of his millions, many people were moved, understandably, to write him letters. His friend Mark Twain wrote him one such: "You seem to be in prosperity. Could you lend an admirer a dollar & a half to buy a hymn book with? God will bless you. I feel it. I know it. . . . P.S. Don't send the hymn-book, send the money."

Among Andrew Carnegie's benefactions was Pittsburgh's Carnegie Institute, with its school (Carnegie Tech), library, museum of natural history, music hall, and art gallery. "This is my monument," he said. By the time he died, it occupied twenty-five acres.

It was a great town to grow up in, Pittsburgh. With one thousand other Pittsburgh schoolchildren, I attended free art classes in Carnegie Music Hall every Saturday morning for four years. Every week, seven or eight chosen kids reproduced their last week's drawings in thick chalks at enormous easels on stage in front of the thousand other kids. After class, everyone scattered; I roamed the enormous building.

Under one roof were the music hall, library, art museum, and natural history museum. Late in the afternoon, after the other kids were all gone, I liked to draw hours-long pencil studies of the chilly marble sculptures in the great hall of classical sculpture. I sat on one man's plinth and drew the next man over—until, during the course of one winter, I had worked my way around the great hall. From these sculptures I learned a great deal about the human leg and not much about the neck, which I could hardly see. I ate a basement-cafeteria lunch and wandered the fabulous building. The natural history museum dominated it.

I felt I was most myself here, here in the churchlike dark lighted by painted dioramas in which tiny shaggy buffalo grazed as far as the eye could see on an enormous prairie I could span with my arms. I could lose myself here, here in the cavernous vault with the shadow of a tyrannosaurus skeleton spread looming all over the domed ceiling, the skeleton shadow enlarged the size of the Milky Way, each bone a dark star.

There was a Van de Graaff generator; you could make a bright crack of lightning strike it from a rod. From a vaulted ceiling hung a cracked wooden skiff—the soul boat of Sesostris III, which Carnegie had picked up in Egypt. Upstairs there were stuffed songbirds in drawers, and empty, faded birdskins in drawers, drab as old handkerchiefs. There were the world's insects on pins and needles; their legs hung down, utterly dead. There were big glass cases you could walk around, in which various motionless American Indians made baskets, started fires, embroidered moccasins, painted pots, chipped spearheads, carried papooses, smoked pipes, drew bows, and skinned rabbits, all of them wearing soft and pale doeskin clothing. The Indians looked stern, even the children, and had bright-red skin. I never thought to draw them; they weren't sculptures.

Sometimes I climbed the broad marble stairs to the art gallery. Carnegie's plans for the art gallery had gone somewhat awry—gang agley—because its first curator was a Scotch-Irish Pittsburgher whose rearing had made it painful for him to spend money. He rarely acquired anything that cost over twenty-five dollars, and liked to buy wee drawings, almost any drawings, in bargain batches, "2 for $10" "3 for $20." By my day, things had improved enormously, and the gallery would buy even large Abstract Expressionist canvases if the artists were guaranteed famous enough. Our school hauled us off to the art gallery once a year for the International Exhibition, but I rarely visited it on my Saturdays in the building, except when *Man Walking* was there.

Carnegie set up the International Exhibition in 1896 to bring contemporary art from all over the world each year to the art museum. Artists competed for a prize, and the museum's curators could buy what they liked, if they felt they could afford it, or if they liked any of it. In 1961, Giacometti's sculpture *Man Walking* won the International. I was sixteen. Everything I knew outside the museum was alien to me, then and for the next few years until I left home.

I saw the sculpture: a wiry, thin person, long legs in full stride, thrust his small, mute head forward into the empty air. Six feet tall, bronze. I read about the sculpture every time I opened the paper; I saw its picture; I climbed the marble stairs alone to look at it again and again. To see *Man Walking,* I walked past abstract canvases by Robert Motherwell, Franz Kline, Adolph Gottlieb. . . . I stopped and looked at their paintings. At school I began to draw abstract forms in rectangles and squares. But more often, then and for many years, I drew what I thought of as the perfect person, whose form matched his inner life, and whose name was, Indian style, Man Walking.

I saw a stilled figure in a swirl of invisible motion. I saw a touchy man moving through a still void. Here was the thinker in the world—but there was no world, only the abyss through which he walked. Man Walking was pure consciousness made poignant: a soul without a culture, absolutely alone, without even a time, without people, speech, books, tools, work, or even clothes. He knew he was walking, here. He knew he was feeling himself walk; he knew he was walking fast and thinking slowly, not forming conclusions, not looking for anything. He himself was barely there. He was in spirit and in form a dissected nerve. He looked freshly made of clay by God, visibly pinched by sure fingertips. He looked like Adam depressed, as if there were no world. He looked like Ahasuerus, condemned to wander without hope. His blind gaze faced the vanishing point.

Man Walking was so skinny his inner life was his outer life; it had nowhere else to go. The point where his head met his spine was the point where spirit met matter. The sculptor's soul floated to his fingertips; I met him there, on Man Walking's skin.

I drew Man Walking in his normal stalking pose and, later, dancing with his arms in the air. What if I fell in love with a man, and he took off his shirt, and I saw he was Man Walking, made of bronze, with Giacometti's thumbprints on him? Well then, I would love him more, for I knew him well; I would hold, if he let me, his twisty head.

Week after week, year after year, after art class I walked the vast museum, and lost myself in the arts, or the sciences. Scientists, it seemed to me as I read the labels on display cases (bivalves, univalves; ungulates, lagomorphs), were collectors and sorters, as I had been. They noticed the things that engaged the curious mind: the way the world develops and divides, colony and polyp, population and tissue, ridge and crystal. Artists, for their part, noticed the things that engaged the mind's private and idiosyncratic interior, that area where the life of the senses mingles with the life of the spirit: the shattering of light into color, and the way it shades off round a bend. The humble attention painters gave to the shadow of a stalk, or the reflected sheen under a chin, or the lapping layers of strong strokes, included and extended the scientists' vision of each least thing as unendingly interesting. But artists laid down the vision in the form of beauty bare—Man Walking—radiant and fierce, inexplicable, and without the math.

It all got noticed: the horse's shoulders pumping; sunlight warping the air over a hot field; the way leaves turn color, brightly, cell by cell; and even the splitting, half-resigned and half-astonished feeling you have when you notice you are walking on earth for a while now—set down for a spell—in this particular time for no particular reason, here.

As a child I read hoping to learn everything, so I could be like my father. I hoped to combine my father's grasp of information and reasoning with my mother's will and vitality. But the books were leading me away. They would propel me right out of Pittsburgh altogether, so I could fashion a life among books somewhere else. So the Midwest nourishes us (Pittsburgh is the Midwest's eastern edge) and presents us with the spectacle of a land and a people completed and certain. And so we run to our bedrooms and read in a fever, and love the big hardwood trees outside the windows, and the terrible Midwest summers, and the terrible Midwest winters, and the forested river valleys with the blue Appalachian Mountains to the east of us and the broad great plains to the west. And so we leave it sorrowfully, having grown strong and restless by opposing with all our will and mind and muscle its simple, loving, single will for us: that we stay, that we stay and find a place among its familiar possibilities. Mother knew we would go; she encouraged us.

I had awakened again, awakened from my drawing and reading, from my exhilarating game playing, from my intense collecting and experimenting, and my cheerful friendships, to see on every side of me a furious procession of which I had been entirely unaware. A procession of fast-talking, keen-eyed, high-stepping, well-dressed men and women of all ages had apparently hoisted me, or shanghaied me, some time ago, and were bearing me breathless along I knew not where. This was the startling world in which I found that I had been living all along. Packed into the procession, I pedaled to keep up, but my feet only rarely hit the ground.

The pace of school life quickened, its bounds tightened, and a new kind of girl emerged from the old. The old-style girl was obedient and tidy. The new-style girl was witty and casual. It was a small school, twenty in a class. We all knew who mattered, not only in our class but in the whole school. The teachers knew, too.

In summer we girls commonly greeted each other, after a perfunctory hello, by extending our forearms side by side to compare tans. We were blond, we were tan, our teeth were white and straightened, our legs were brown and depilated, our blue eyes glittered pale in our dark faces; we laughed; we shuffled the cards fast and dealt four hands. It was not for me. I hated it so passionately I thought my shoulders and arms, swinging at the world, would split off from my body like loose spinning blades, and fly wild and slice everyone up. With all my heart, sometimes, I longed for the fabled Lower East Side of Manhattan, for Brooklyn, for the Bronx, where the thoughtful and feeling people in books grew up on porch stoops among seamstress intellectuals. There I belonged if anywhere, there where the book people were—recent Jewish immigrants, everybody deep every livelong minute. I could just see them, sitting there feeling deeply. Here, instead, I saw polished fingernails clicking, rings flashing, gold bangle bracelets banging and ringing together as sixteen-year-old girls like me pushed their cuticles back, as they ran combs through their just-washed, just-cut, just-set hair, as they lighted Marlboros with hard snaps of heavy lighters, and talked about other girls or hair. It never crossed my mind that you can't guess people's lives from their chatter.

This was the known world. Women volunteered, organized the households, and reared the kids; they kept the traditions, and taught by example a dozen kinds of love. Mother polished the brass, wiped the ashtrays, stood barefoot on the couch to hang a picture. Margaret Butler washed the windows, which seemed to yelp. Mother dusted and polished the big philodendrons, tenderly, leaf by leaf, as if she were washing babies' faces. Margaret came sighing down the stairs with an armful of laundry or wastebaskets. Mother inspected the linens for a party; she fetched from a closet the folding felted boards she laid over the table. Margaret turned on the vacuum cleaner again. Mother and Margaret changed the sheets and pillowcases.

Then Margaret left. I had taken by then to following her from room to room, trying to get her to spill the beans about being black; she kept moving. Nothing changed. Mother wiped the stove; she ran the household with her back to it. You heard a staccato in her voice, and saw the firm force of her elbow, as she pressed hard on a dried tan dot of bean soup, and finally took a fingernail to it, while quizzing Amy about a car pool to dancing school, and me about a ride back from a game. No page of any book described housework, and no one mentioned it; it didn't exist. There was no such thing.

A woman at our country club, a prominent figure at our church, whose daughters went to Ellis, never washed her face all summer, to preserve her tan. We rarely saw the pale men at all; they were off pulling down the money on which the whole scene floated. Most men came home exhausted in their gray suits to scantily clad women smelling of Bain de Soleil, and do-nothing tanned kids in Madras shorts.

There was real beauty to the old idea of living and dying where you were born. You could hold a place in a kind of eternity. Your grandparents took you out to dinner Sunday nights at the country club, and you could take your own grandchildren there when that time came: more little towheads, as squint-eyed and bony-legged and Scotch-Irish as hillbillies. And those grandchildren, like figures in a reel endlessly unreeling, would partake of the same timeless, hushed, muffled sensations.

They would join the buffet line on Sunday nights in winter at the country club. I remember: the club lounges before dinner dimly lighted and opulent like the church; the wool rugs absorbing footsteps; the lined damask curtains lapping thickly across tall, leaded-glass windows. The adults drank old-fashioneds. The fresh-haired children subsisted on bourbon-soaked maraschino cherries, orange slices, and ice cubes. They roved the long club corridors in slippery shoes; they opened closet doors, tried to get outside, laughed so hard they spit their ice cubes, and made sufficient commotion to rouse the adults to dinner. In the big dining room, layers of fine old unstarched linen draped the tables as thickly as hospital beds. Heavy-bottomed glasses sank into the tablecloths soundlessly.

And sempiternal too were the summer dinners at the country club, the sun-shocked people somnambulistic as angels. The children's grandchildren could see it. Space and light multiplied the club rooms; the damask curtains were heaved back; the French doors now gave out onto a flagstone terrace overlooking the swimming pool, near the sixth hole. On the terrace, men and women drank frozen daiquiris, or the unvarying Scotch, and their crystal glasses clicked on the glass tabletops, and then stuck in pools of condensation as if held magnetically, so they had to skid the glasses across the screeching tabletops to the edges in order to raise them at all. The cast-iron chair legs, painted white, marked and chipped the old flagstones, and dug up the interstitial grass.

The dressed children on the terrace looked with longing down on the tanned and hilarious children below. The children below wouldn't leave the pool, although it was seven-thirty; they knew no parent would actually shout at them from the flagstone terrace above. When these poolside children jumped in the water, the children on the terrace above could see their shimmering gray bodies against the blue pool. The water knit a fabric of light over their lively torsos and limbs, a loose gold chain mail. They looked like fish swimming in wide gold nets.

The children above were sunburnt, and their cotton dresses scraped their shoulders. The outsides of their skins felt hot, and the insides felt cold, and they tried to warm one arm with another. In summer, no one drank old-fashioneds, so there was nothing for children to eat till dinner.

This was the world we knew best—this, and Oma's.

Oma's world was no likely alternative to ours; Oma had a chauffeur and her chauffeur had to drink from his own glass.

My forays into Oma's world changed. I was working in the summers now. The summer I sold men's bathing suits, I ate lunch alone in a dark bar and played the numbers for a quarter every week, right there in the underworld. I no longer went to the Lake with Amy. But for a few spring vacations after our grandfather died, Amy and I visited Oma and Mary in their apartment in Pompano Beach, Florida.

On my last visit, I was fifteen. Everything I was required to do, such as sit at a table with other people, either bored me to fury or infuriated me to a kind of benumbed lethargy. I was finding it difficult to live—finished with everything I knew and ignorant of anything else. I woke every morning full of hope, and was livid with rage before breakfast, at one thing or another.

Oma and I argued that year, over a word. Because something I was talking about seemed to require it, Oma said the word for padded, upholstered furniture was "overstuffed." I wouldn't hear of it, having never heard of it. "It's not overstuffed; it's stuffed just right." Oma pointed out that it was just barely possible that she knew something on earth that I didn't. I couldn't quite believe her.

In Oma's Pompano Beach apartment, I lounged on the bright print bamboo furniture and looked at the Asian objects she had been collecting all her life: gaudy Chinese cloisonné lamps, lacquered chests, sentimental Japanese porcelain figurines—women in whiteface with cocked heads and pink circles on their cheeks—gold, bossed mirrors, foot-long yellow ashtrays shaped like carp, and a pair of green ceramic long-tailed birds, which took up the breakfast table. It was years before I learned that Asian art was supposed to be delicate.

In Florida, Mary Burinda drove the machine. Oma rode in the front seat; Amy and I sat in back. That year, Oma's current, roseate Cadillac had an extra row of upholstered seats, which folded against the front seat's back—like, but not very like, the extra seats in a cab. An especially long distance stretched between the front seat and the back.

One day, we were driving back from Miami; Oma had been "looking at shoes." (Oma had announced at breakfast, "Today I want to look at shoes," and I repeated the phrase to myself all morning, marveling, to learn what it might feel like to want to look at shoes.)

Without provocation, she broke down, grieving for our grandfather. She rubbed her round face in her hands. Mary, at the wheel, expostulated, shocked, "*Missus* Doak. Oh, Missus *Doak*." She added, "That was two years ago, Missus Doak." This occasioned a fresh outburst, which broke our hearts. I saw Oma's red hair and her lowered head wipe back and forth.

Then she rallied and began defensively, "But you know, he was never cross with me."

"Never once?" someone ventured from the depths of the back seat.

"Well, once. Yes, once." Her voice lightened.

They were driving, she said, on a high mountain road. I saw the back of her round head swivel; she was looking up and away, remembering. The two of them were driving along a dreadful road, she said, a perfectly horrible road, in Tennessee maybe. Her voice grew shrill.

"There was a sheer drop just outside my window, and I thought we were going over. We were going over, I tell you." She was furious at the thought. "And he got very cross with me."

She had never seen him so angry. "He said I could either hush, or get out and walk. Can you imagine!"

She was awed. So was I. We were both awed, that he had dared. It cheered everyone right up.

The bird-watching was fine in the nearby Fort Lauderdale city park. Right in the middle of town, the park was mostly wild forest, with a few clearings and roads. Oma and Mary drove me to the park early every morning, and picked me up at noon. There I saw some of the few smooth-billed anis in the United States. They were black parrot-beaked birds; they hung around the park's dump. The binoculars I wore banged against my skinny rib cage. I filled a notebook with sketches, information, and records. I saw myrtle warblers in the clearings. I saw a coot and a purple gallinule side by side, just as Peterson had painted them in the field guide; they swam in a lagoon under sea grape trees. They seemed, as common birds seem to the delirious beginner, miraculous and rare. (The tizzy that birds excite in the beginner are a property of the beginner, not of the birds; so those who love the tizzy itself must ever keep beginning things.)

Often I was startled to see, through binoculars and flattened by their lenses, glimpsed through the dark subtropical leaves, the white hull of some pleasure cruiser setting out on a Lauderdale canal. Who would go cruising beside houses and lawns, when he could be watching smooth-billed anis? I alone was sane, I thought, in a world of crazy people. Standing in the park's smelly dump, I shrugged.

Afternoons I wandered the blinding beach, swam, and read about tide pools in Maine; I was reading *The Edge of the Sea.* On the beach I found skeletons of velella, or by-the-wind-sailors. From the high apartment windows I looked at the lifeguards around the pool below, and wondered how I might meet them. By day, Oma and Mary shopped. Evenings we went out to dinner. Amy was as desperately bored as I was, but I wouldn't let her follow me; I addressed her in French. Everyone knew this was our last Florida trip.

It was on this visit that Oma asked me, when we were alone, what exactly it was that homosexuals did. She was miffed that she'd been unable to command this information before now. She said she'd wondered for many years without knowing who she could ask.

Amy and I boarded our plane back to Pittsburgh. It would be softball season at school, and a new baseball season for the Pirates, whose hopes were resting on a left-handed reliever, Elroy Face, and on the sober starter, Vernon Law—the Deacon—and on the big bat of our right fielder, Roberto Clemente, whom everyone in town adored.

Flying back, looking out over the Blue Ridge, I remembered a game I had seen at Forbes Field the year before: Clemente had thrown from right field to the plate, as apparently easily as a wheel spins. The ball seemed not to arc at all; the throw caught the runner from third. You could watch this man at inning's end lope from right field to the dugout, and you'd weep—at the way his joints moved, and the ease and power in his spine.

I was ready for all that, but it was only late March, and snowing in Pittsburgh when we got off the plane, and dark. At least we were tan.

When i was fifteen, i felt it coming; now I was sixteen, and it hit.

My feet had imperceptibly been set on a new path, a fast path into a long tunnel like those many turnpike tunnels near Pittsburgh, turnpike tunnels whose entrances bear on brass plaques a roll call of those men who died blasting them. I wandered witlessly forward and found myself going down, and saw the light dimming; I adjusted to the slant and dimness, traveled further down, adjusted to greater dimness, and so on. There wasn't a whole lot I could do about it, or about anything. I was going to hell on a handcart, that was all, and I knew it and everyone around me knew it, and there it was.

I was growing and thinning, as if pulled. I was getting angry, as if pushed. I morally disapproved most things in North America, and blamed my innocent parents for them. My feelings deepened and lingered. The swift moods of early childhood—each formed by and suited to its occasion—vanished. Now feelings lasted so long they left stains. They arose from nowhere, like winds or waves, and battered at me or engulfed me.

When I was angry, I felt myself coiled and longing to kill someone or bomb something big. Trying to appease myself, during one winter I whipped my bed every afternoon with my uniform belt. I despised the spectacle I made in my own eyes—whipping the bed with a belt, like a creature demented!—and I often began halfheartedly, but I did it daily after school as a desperate discipline, trying to rid myself and the innocent world of my wildness. It was like trying to beat back the ocean.

Sometimes in class I couldn't stop laughing; things were too funny to be borne. It began then, my surprise that no one else saw what was so funny.

I read some few books with such reverence I didn't close them at the finish, but only moved the pile of pages back to the start, without breathing, and began again. I read one such book, an enormous novel, six times that way—closing the binding between sessions, but not between readings.

On the piano in the basement I played the maniacal "Poet and Peasant Overture" so loudly, for so many hours, night after night, I damaged the piano's keys and strings. When I wasn't playing this crashing overture, I played boogie-woogie, or something else, anything else, in octaves—otherwise, it wasn't loud enough. My fingers were so strong I could do push-ups with them. I played one piece with my fists. I banged on a steel-stringed guitar till I bled, and once on a particularly piercing rock-and-roll downbeat I broke straight through one of Father's snare drums.

I loved my boyfriend so tenderly, I thought I must transmogrify into vapor. It would take spectroscopic analysis to locate my molecules in thin air. No possible way of holding him was close enough. Nothing could cure this bad case of gentleness except, perhaps, violence: maybe if he swung me by the legs and split my skull on a tree? Would that ease this insane wish to kiss too much his eyelids' outer corners and his temples, as if I could love up his brain?

I envied people in books who swooned. For two years I felt myself continuously swooning and continuously unable to swoon; the blood drained from my face and eyes and flooded my heart; my hands emptied, my knees unstrung, I bit at the air for something worth breathing—but I failed to fall, and I couldn't find the way to black out. I had to live on the lip of a waterfall, exhausted.

When I was bored I was first hungry, then nauseated, then furious and weak. "Calm yourself," people had been saying to me all my life. Since early childhood I had tried one thing and then another to calm myself, on those few occasions when I truly wanted to. Eating helped; singing helped. Now sometimes I truly wanted to calm myself. I couldn't lower my shoulders; they seemed to wrap around my ears. I couldn't lower my voice although I could see the people around me flinch. I waved my arm in class till the very teachers wanted to kill me.

I was what they called a live wire. I was shooting out sparks that were digging a pit around me, and I was sinking into that pit. Laughing with Ellin at school recess, or driving around after school with Judy in her jeep, exultant, or dancing with my boyfriend to Louis Armstrong across a polished dining-room floor, I got so excited I looked around wildly for aid; I didn't know where I should go or what I should do with myself. People in books split wood.

When rage or boredom reappeared, each seemed never to have left. Each so filled me with so many years' intolerable accumulation it jammed the space behind my eyes, so I couldn't see. There was no room left even on my surface to live. My rib cage was so taut I couldn't breathe. Every cubic centimeter of atmosphere above my shoulders and head was heaped with last straws. Black hatred clogged my very blood. I couldn't peep, I couldn't wiggle or blink; my blood was too mad to flow.

For as long as I could remember, I had been transparent to myself, unselfconscious, learning, doing, most of every day. Now I was in my own way; I myself was a dark object I could not ignore. I couldn't remember how to forget myself. I didn't want to think about myself, to reckon myself in, to deal with myself every livelong minute on top of everything else—but swerve as I might, I couldn't avoid it. I was a boulder blocking my own path. I was a dog barking between my own ears, a barking dog who wouldn't hush.

So this was adolescence. Is this how the people around me had died on their feet—inevitably, helplessly? Perhaps their own selves eclipsed the sun for so many years the world shriveled around them, and when at last their inescapable orbits had passed through these dark egoistic years it was too late, they had adjusted.

Must I then lose the world forever, that I had so loved? Was it all, the whole bright and various planet, where I had been so ardent about finding myself alive, only a passion peculiar to children, that I would outgrow even against my will?

I QUIT THE CHURCH. I wrote the minister a fierce letter. The assistant minister, kindly Dr. James H. Blackwood, called me for an appointment. My mother happened to take the call.

"Why," she asked, "would he be calling you?" I was in the kitchen after school. Mother was leaning against the pantry door, drying a crystal bowl.

"What, Mama? Oh. Probably," I said, "because I wrote him a letter and quit the church."

"You—what?" She began to slither down the doorway, weak-kneed, like Lucille Ball. I believe her whole life passed before her eyes.

As I climbed the stairs after dinner I heard her moan to Father, "She wrote the minister a letter and quit the church."

"She—what?"

Father knocked on the door of my room. I was the only person in the house with her own room. Father ducked under the doorway, entered, and put his hands in his khakis' pockets. "Hi, Daddy." Actually, it drove me nuts when people came in my room. Mother had come in just last week. My room was getting to be quite the public arena. Pretty soon they'd put it on the streetcar routes. Why not hold the U.S. Open here? I was on the bed, in uniform, trying to read a book. I sat up and folded my hands in my lap.

I knew that Mother had made him come—"She listens to you." He had undoubtedly been trying to read a book, too.

Father looked around, but there wasn't much to see. My

rock collection was no longer in evidence. A framed tiger swallowtail, spread and only slightly askew on white cotton, hung on a yellowish wall. On the mirror I'd taped a pencil portrait of Rupert Brooke; he was looking off softly. Balanced on top of the mirror were some yellow-and-black FALLOUT SHELTER signs, big aluminum ones, which Judy had collected as part of her antiwar effort. On the pale maple desk there were, among other books and papers, an orange thesaurus, a blue three-ring binder with a boy's name written all over it in every typeface, a green assignment notebook, and Emerson's *Essays*.

Father began, with some vigor: "What was it you said in this brilliant letter?" He went on: But didn't I see? That people did these things—quietly? Just—quietly? No fuss? No flamboyant gestures. No uncalled-for letters. He was forced to conclude that I was deliberately setting out to humiliate Mother and him.

"And your poor sisters, too!" Mother added feelingly from the hall outside my closed door. She must have been passing at that very moment. Then, immediately, we all heard a hideous shriek ending in a wail; it came from my sisters' bathroom. Had Molly cut off her head? It set us all back a moment—me on the bed, Father standing by my desk, Mother outside the closed door—until we all realized it was Amy, mad at her hair. Like me, she was undergoing a trying period, years long; she, on her part, was mad at her hair. She screeched at it in the mirror; the sound carried all over the house, kitchen, attic, basement, everywhere, and terrified all the rest of us, every time.

The assistant minister of the Shadyside Presbyterian Church, Dr. Blackwood, and I had a cordial meeting in his office. He was an experienced, calm man in a three-piece suit; he had a mustache and wore glasses. After he asked me why I had quit the church, he loaned me four volumes of C. S. Lewis's broadcast talks, for a paper I was writing. Among the volumes proved to be *The Problem of Pain,* which I would find fascinating, not quite serious enough, and too short. I had already written a paper on the Book of Job. The subject scarcely seemed to be closed. If the all-powerful creator directs the world, then why all this suffering? Why did the innocents die in the camps, and why do they starve in the cities and farms? Addressing this question, I found thirty pages written thousands of years ago, and forty pages written in 1955. They offered a choice of fancy language saying, "Forget it," or serenely worded, logical-sounding answers that so strained credibility (pain is God's megaphone) that "Forget it" seemed in comparison a fine answer. I liked, however, C. S. Lewis's effort to defuse the question. The sum of human suffering we needn't worry about: There is plenty of suffering, but no one ever suffers the sum of it.

Dr. Blackwood and I shook hands as I left his office with his books.

"This is rather early of you, to be quitting the church," he said as if to himself, looking off, and went on mildly, almost inaudibly, "I suppose you'll be back soon."

Humph, I thought. Pshaw.

Now it was May. Daylight Saving Time had begun; the colored light of the long evenings fairly split me with joy. White trillium had bloomed and gone on the forested slopes in Fox Chapel. The cliffside and riverside patches of woods all over town showed translucent ovals of yellow or ashy greens; the neighborhood trees on Glen Arden Drive had blossomed in white and red.

Baseball season had begun, a season which recalled but could never match last year's National League pennant and seventh-game World Series victory over the Yankees, when we at school had been so frenzied for so many weeks they finally and wisely opened the doors and let us go. I had walked home from school one day during that series and seen Pittsburgh's Fifth Avenue emptied of cars, as if the world were over.

A year of wild feelings had passed, and more were coming. Without my noticing, the drummer had upped the tempo. Someone must have slipped him a signal when I wasn't looking; he'd speeded things up. The key was higher, too. I had a driver's license. When I drove around in Mother's old Dodge convertible, the whole town smelled good. And I did drive around the whole town. I cruised along the blue rivers and across them on steel bridges, and steered up and down the scented hills. I drove winding into and out of the steep neighborhoods across the Allegheny River, neighborhoods where I tried in vain to determine in what languages the signs on storefronts were written. I drove onto boulevards, highways, beltways, freeways, and the turnpike. I could drive to Guatemala, drive to Alaska. Why, I asked myself, did I drive to—of all spots on earth—our garage? Why home, why school?

Throughout the long, deadly school afternoons, we junior and senior girls took our places in study hall. We sat at desks in a roomful of desks, whether or not we had something to do, until four o'clock.

Now this May afternoon a teacher propped open the study hall's back door. The door gave onto our hockey field and, behind it, Pittsburgh's Nabisco plant, whence, O Lordy, issued the smell of shortbread today; they were baking Lorna Doones. Around me sat forty or fifty girls in green cotton jumpers and spring-uniform white bucks. They rested their chins on the heels of both hands and leaned their cheeks on curled fingers; their propped heads faced the opened pages of *L'Étranger, Hamlet, Vanity Fair*. Some girls leaned back and filed their nails. Some twisted stiff pieces of their hair, to stay not so much awake as alive. Sometimes in health class, when we were younger, we had all been so bored we hooked our armpits over our chairs' backs so we cut off all circulation to one arm, in an effort to kill that arm for something to do, or cause a heart attack, whichever came first. We were, in fact, getting a dandy education. But sometimes we were restless. Weren't there some wars being fought somewhere that I, for one, could join?

I wrote a name on a notebook. I looked at the study-hall ceiling and tried to see that boy's familiar face—light and dark, bold-eyed, full of feeling—on the inside of my eyelids. Failing that, I searched for his image down the long speckled tunnel or corridor I saw with my eyes closed. As if visual memory were a Marx brothers comedy, I glimpsed swift fragments—a wry corner of his lip, a pointy knuckle, a cupped temple—which crossed the corridor so fast I recognized them only as soon as they vanished. I opened my eyes and wrote his name. His depth and complexity were apparently infinite. From the tip of his lively line of patter to the bottom of his heartbroken, hopeful soul was the longest route I knew, and the best.

The heavy, edible scent of shortbread maddened me in my seat, made me so helpless with longing my wrists gave out; I couldn't hold a pen. I looked around constantly to catch someone's eye, anyone's eye.

It was a provocative fact, which I seemed to have discovered, that we students outnumbered our teachers. Must we then huddle here like sheep? By what right, exactly, did these few women keep us sitting here in this clean, bare room to no purpose? Lately I had been trying to enflame my friends with the implications of our greater numbers. We could pull off a riot. We could bang on the desks and shout till they let us out. Then we could go home and wait for dinner. Or we could bear our teachers off on our shoulders, and—what? Throw them into the Lorna Doone batter? I got no takers.

I had finished my work long ago. "Works only on what interests her," the accusation ran—as if, I reflected, obedience outranked passion, as if sensible people didn't care what they stuck in their minds. Today as usual no one around me was ready for action. I took a fresh sheet of paper and copied on it random lines in French:

> *Ô saisons, ô châteaux!*
> Is it through these endless nights that you sleep
> in exile
> Ô million golden birds, ô future vigor?
>
> Oh, that my keel would split! Oh, that I would
> go down in the sea!

I had struck upon the French Symbolists, like a canyon of sharp crystals underground, like a long and winding corridor lined with treasure. These poets popped into my ken in an odd way: I found them in a book I had rented from a drugstore. Carnegie and school libraries filled me in. I read Enid Starkie's Rimbaud biography. I saved my allowance for months and bought two paperbound poetry books, the Penguin *Rimbaud,* and a Symbolist anthology in which Paul Valéry declaimed, "*Azure! c'est moi . . .*" I admired Gérard de Nerval. This mad writer kept a lobster as a pet. He walked it on a leash along the sidewalks of Paris, saying, "It doesn't bark, and knows the secrets of the deep."

I loved Rimbaud, who ran away, loved his skinny, furious face with the wild hair and snaky, unseeing eyes pointing in two directions, and his poems' confusion and vagueness, their overwritten longing, their hatred, their sky-shot lyricism, and their oracular fragmentation, which I enhanced for myself by reading and retaining his stuff in crazed bits, mostly from *Le Bateau Ivre,* The Drunken Boat. (The drunken boat tells its own story, a downhill, down-stream epic unusually full of words.)

Now in study hall I saw that I had drawn all over this page; I got out another piece of paper. Rimbaud was damned. He said so himself. Where could I meet someone like that? I wrote down another part:

> There is a cathedral that goes down and a lake
> that goes up.
> There is a troupe of strolling players in costume,
> glimpsed on the road through the edge of
> the trees.

I looked up from the new page I had already started to draw all over. Except for my boyfriend, the boys I knew best were out of town. They were older, prep-school and college boys whose boldness, wit, breadth of knowledge, and absence of scruples fascinated me. They cruised the deb party circuit all over Pennsylvania, holding ever-younger girls up to the light like chocolates, to determine how rich their centers might be. I smiled to recall one of these boys: he was so accustomed to the glitter of society, and so sardonic and graceful, that he carried with him at all times, in his jacket pocket, a canister of dance wax. Ordinary boys carried pocket knives for those occasions which occur unexpectedly, and this big, dark-haired boy carried dance wax for the same reason. When the impulse rose, he could simply sprinkle dance wax on any hall or dining-room floor, take a girl in his arms, and whirl her away. I had known these witty, handsome boys for years, and only recently understood that when they were alone, they read books. In public, they were lounge lizards; they drank; they played word games, filling in the blanks desultorily; they cracked wise. These boys would be back in town soon, and my boyfriend and I would join them.

Whose eye could I catch? Everyone in the room was bent over her desk. Ellin Hahn was usually ready to laugh, but now she was working on something. She would call me as soon as we got home. Every day on the phone, I unwittingly asked Ellin some blunt question about the social world around us, and at every question she sighed and said to me, "You still don't get it"—or often, as if addressing a jury of our incredulous peers, "She still doesn't get it!"

Looking at the study-hall ceiling, I dosed myself almost fatally with the oxygen-eating lines of Verlaine's "The long sobs / of the violins / of autumn / wound my heart / with a languor / monotone."

This unsatisfying bit of verse I repeated to myself for ten or fifteen minutes, by the big clock, over and over, clobbering myself with it, the way Molly, when she had been a baby, banged the top of her head on her crib.

Ô world, ô college, ô dinner . . .
Ô unthinkable task . . .

Funny how badly I'd turned out. Now I was always in trouble. It felt as if I was doing just as I'd always done—I explored the neighborhood, turning over rocks. The latest rocks were difficult. I'd been in a drag race, of all things, the previous September, and in the subsequent collision, and in the hospital; my parents saw my name in the newspapers, and their own names in the newspapers. Some boys I barely knew had cruised by that hot night and said to a clump of us girls on the sidewalk, "Anybody want to come along for a drag race?" I did, absolutely. I loved fast driving.

It was then, in the days after the drag race, that I noticed the ground spinning beneath me, all bearings lost, and recognized as well that I had been loose like this—detached from all I saw and knowing nothing else—for months, maybe years. I whirled through the air like a bull-roarer spun by a lunatic who'd found his rhythm. The pressure almost split my skin. What else can you risk with all your might but your life? Only a moment ago I was climbing my swing set, holding one cold metal leg between my two legs tight, and feeling a piercing oddness run the length of my gut—the same sensation that plucked me when my tongue touched tarnish on a silver spoon. Only a moment ago I was gluing squares of paper to rocks; I leaned over the bedroom desk. I was drawing my baseball mitt in the attic, under the plaster-stain ship; a pencil study took all Saturday morning. I was capturing the flag, turning the double play, chasing butterflies by the country-club pool. Throughout these many years of childhood, a transparent sphere of timelessness contained all my running and spinning as a glass paperweight holds flying snow. The sphere of this idyll broke; time unrolled before me in a line. I woke up and found myself in juvenile court. I was hanging from crutches; for a few weeks after the drag race, neither knee worked. (No one else got hurt.) In juvenile court, a policeman wet all ten of my fingertips on an ink pad and pressed them, one by one, using his own fingertips, on a form for the files.

Turning to the French is a form of suicide for the American who loves literature—or, as the joke might go, it is at least a cry for help. Now, when I was sixteen, I had turned to the French. I flung myself into poetry as into Niagara Falls. Beauty took away my breath. I twined away; I flew off with my eyes rolled up; I dove down and succumbed. I bought myself a plot in Valéry's marine cemetery, and moved in: cool dirt on my eyes, my brain smooth as a cannonball. It grieves me to report that I tried to see myself as a sobbing fountain, apparently serene, tall and thin among the chill marble monuments of the dead. Rimbaud wrote a lyric that gently described a man sleeping out in the grass; the sleeper made a peaceful picture, until, in the poem's last line, we discover in his right side two red holes. This, and many another literary false note, appealed to me.

I'd been suspended from school for smoking cigarettes. That was a month earlier, in early spring. Both my parents wept. Amy saw them weeping; horrified, she began to cry herself. Molly cried. She was six, missing her front teeth. Like Mother and me, she had pale skin that turned turgid and red when she cried; she looked as if she were dying of wounds. I didn't cry, because, actually, I was an intercontinental ballistic missile, with an atomic warhead; they don't cry.

Why didn't I settle down, straighten out, shape up? I wondered, too. I thought that joy was a childish condition that had forever departed; I had no glimpse then of its return the minute I got to college. I couldn't foresee the pleasure—or the possibility—of shedding sophistication, walking away from rage, and renouncing French poets.

While I was suspended from school, my parents grounded me. During that time, Amy began to visit me in my room.

When she was thirteen, Amy's beauty had grown inconspicuous; she seemed merely pleasant-looking and tidy. Her green uniform jumper fit her neatly; her thick hair was smoothly turned under; her white McMullen collars looked sweet. She had a good eye for the right things; people respected her for it. I think that only we at home knew how spirited she could get. "Oh, no!" she cried when she laughed hard. "Oh, no!" Amy adored our father, rather as we all did, from afar. She liked boys whose eyebrows met over their noses. She liked boys, emphatically; she followed boys with her big eyes, awed.

In my room, Amy listened to me rant; she reported her grade's daily gossip, laughed at my jokes, cried, "Oh, no!" and told me about the book she was reading, Wilkie Collins, *The Woman in White*. I liked people to tell me about the books they were reading. Next year, Amy was going to boarding school in Philadelphia; Mother had no intention of subjecting the family to two adolescent maelstroms whirling at once in the same house.

Late one night, my parents and I sat at the kitchen table; there was a truce. We were all helpless, and tired of fighting. Amy and Molly were asleep.

"What are we going to do with you?"

Mother raised the question. Her voice trembled and rose with emotion. She couldn't sit still; she kept getting up and roaming around the kitchen. Father stuck out his chin and rubbed it with his big hands. I covered my eyes. Mother squeezed white lotion into her hands, over and over. We all smoked; the ashtray was full. Mother walked over to the sink, poured herself some ginger ale, ran both hands through her short blond hair to keep it back, and shook her head.

She sighed and said again, looking up and out of the night-black window, "Dear God, what are we going to do with you?" My heart went out to them. We all seemed to have exhausted our options. They asked me for fresh ideas, but I had none. I racked my brain, but couldn't come up with anything. The U.S. Marines didn't take sixteen-year-old girls.

Outside the study hall that May, a cardinal sang his round-noted song, and a robin sang his burbling song, and I slumped at my desk with my heart pounding, too harried by restlessness to breathe. I collected poems and learned them. I found the British war poets—World War I: Rupert Brooke, Edmund Blunden, Siegfried Sassoon, and especially Wilfred Owen, who wrote bitterly without descending to sarcasm. I found Asian and Middle Eastern poetry in translation—whole heaps of lyrics fierce or limp—which I ripped to fragments for my collection. I wanted beauty bare of import; I liked language in strips like pennants.

Under the spell of Rimbaud I wrote a poem that began with a line from *Une Saison en Enfer,* "Once, if I remember well," and continued, "My flesh did lie confined in hell." It ended, slantingly, to my own admiration, "And in my filth did I lie still." I wrote other poems, luscious ones, in the manner of the Song of Songs. One teacher, Miss Hickman, gave her lunch hour to meet with us about our poems.

It galled me that adults, as a class, approved the writing and memorization of poetry. Wasn't poetry secret and subversive? One sort of poetry was full of beauty and longing; it exhaled, enervated and helpless, like Li Po. Other poems were threats and vows. They inhaled; they poured into me a power I could not spend. The best of these, a mounted Arabic battle cry, I recited to myself by the hour, hoping to trammel the teachers' drone with hoofbeats.

I dosed myself with pure lyricism; I lived drugged on sensation, as I had lived alert on sensation as a little child. I wanted to raise armies, make love to armies, conquer armies. I wanted to swim in the stream of beautiful syllables until I tired. I wanted to bust up the Ellis School with my fists.

One afternoon at Judy Schoyer's house, I saw a white paperback book on a living-room chair: Lucretius, *On the Nature of Things*. Lucretius, said the book's back cover, had flourished in the first century B.C. This book was a prose translation of a long poem in Latin hexameters, the content of which was ancient physics mixed with philosophy. Why was this book in print? Why would anyone read wrong science, the babblings of a poet in a toga—why but from disinterested intellectual curiosity? I regarded the white paperback book as if it had been a meteorite smoldering on the chair's silk upholstery.

It was Judy's father's book. Mr. Schoyer loaned me the book when he was finished with it, and I read it; it was deadly dull. Nevertheless, I admired Judy's lawyer father boundlessly. I could believe in him for months at a time. His recreation proceeded from book to book, and had done so all his life. He had, I recalled, majored in classical history and literature. He wanted to learn the nature of things. He read and memorized poetry. He quizzed us about current events—what is your opinion of our new Supreme Court justice? On the other hand, his mother's family were Holyokes, and he hadn't raised a hand to rescue Judy from having to come out in Salem, Massachusetts. She had already done so, and would not talk about it.

Judy was tall now, high-waisted, graceful, messy still; she smiled forgivingly, smiled ironically, behind her thick glasses. Her limbs were thin as stalks, and her head was round. She spoke softly. She laughed at anything chaotic. Her family took me to the ballet, to the Pittsburgh Symphony, to the Three Rivers Arts Festival; they took me ice skating on a frozen lake in Highland Park, and swimming in Ohiopyle, south of town where the Youghiogheny River widens over flat rock outcrops.

After school, we piled in Judy's jeep. Out of the jeep's open back I liked to poke the long barrel of a popgun, slowly, and aim it at the drivers of the cars behind us, and shoot the cork, which then swung from its string. The drivers put up their hands in mock alarm, or slumped obligingly over their wheels. Pittsburghers were wonderful sports.

All spring long I crawled on my pin. I was reading *General Semantics*—Alfred Korzybski's early stab at linguistics; I'd hit on it by accident, in books with the word "language" in their titles. I read Freud's standard works, which interested me at first, but they denied reason. Denying reason had gotten Rimbaud nowhere. I read without snobbery, excited and alone, wholly free in the indifference of society. I read with the pure, exhilarating greed of readers sixteen, seventeen years old; I felt I was exhuming lost continents and plundering their stores. I knocked open everything in sight—Henry Miller, Helen Keller, Hardy, Updike, and the French. The war novels kept coming out, and so did John O'Hara's. I read popular social criticism with Judy and Ellin—*The Ugly American, The Hidden Persuaders, The Status Seekers.* I thought social and political criticism were interesting, but not nearly so interesting as almost everything else.

Ralph Waldo Emerson, for example, excited me enormously. Emerson was my first crack at Platonism, Platonism as it had come bumping and skidding down the centuries and across the ocean to Concord, Massachusetts. Emerson was a thinker, full time, as Pasteur and Salk were full-time biologists. I wrote a paper on Emerson's notion of the soul—the oversoul, which, if I could banish from my mind the thought of galoshes (one big galosh, in which we have our being), was grand stuff. It was metaphysics at last, poetry with import, philosophy minus the Bible. And Emerson incited to riot, flouting every authority, and requiring each native to cobble up an original relation with the universe. Since rioting seemed to be my specialty, if only by default, Emerson gave me heart.

Enervated, fanatic, filled long past bursting with oxygen I couldn't use, I hunched skinny in the school's green uniform, etiolated, broken, bellicose, starved, over the back-breaking desk. I sighed and sighed but never emptied my lungs. I said to myself, "O breeze of spring, since I dare not know you, / Why part the silk curtains by my bed?" I stuffed my skull with poems' invisible syllables. If unauthorized persons looked at me, I hoped they'd see blank eyes.

On one of these May mornings, the school's headmistress called me in and read aloud my teachers' confidential appraisals. Madame Owens wrote an odd thing. Madame Owens was a sturdy, affectionate, and humorous woman who had lived through two world wars in Paris by eating rats. She had curly black hair, rouged cheeks, and long, sharp teeth. She swathed her enormous body in thin black fabrics; she sat at her desk with her tiny ankles crossed. She chatted with us; she reminisced.

Madame Owens's kind word on my behalf made no sense. The headmistress read it to me in her office. The statement began, unforgettably, "Here, alas, is a child of the twentieth century." The headmistress, Marion Hamilton, was a brilliant and strong woman whom I liked and respected; the school's small-minded trustees would soon run her out of town on a rail. Her black hair flared from her high forehead. She looked up at me significantly, raising an eyebrow, and repeated it: "Here, alas, is a child of the twentieth century."

I didn't know what to make of it. I didn't know what to do about it. You got a lot of individual attention at a private school.

My idea was to stay barely alive, pumping blood and exchanging gases just enough to sustain life—but certainly not enough so that anyone suspected me of sentience, certainly not enough so that I woke up and remembered anything—until the time came when I could go.

> *C'est elle, la petite morte, derrière les rosiers . . .*
> It is she, the little dead girl, behind the rose bushes . . .
> the child left on the jetty washed out to sea,
> the little farm child following the lane
> whose forehead touches the sky.

During classes all morning, I drew. Drawing deliberately, as I had learned to do, yielded complex, fresh drawings: the inevitable backs of my friends' heads; their ankles limp at rest over their winter brown oxfords; the way their white shirts' shoulders emerged from their uniform jumpers. I roused myself to these efforts only once or twice a day. I drew Man Walking, too. During the other six or seven hours, when I wasn't fiddling with poetry, I drew at random.

Drawing at random, paying no attention, infuriated me, yet I never stopped. For years as a child I drew faces on the back of my left hand, on the tops of my knees, in my green assignment book, my blue canvas three-ring binder. Later I drew rigid faces on the Latin textbook's mazy printed page, down and across the spaces between lines and words. I drew stretchable cartoons on the wiggly and problematic plane of a book's page edges. Those page edges—pressed slats and slits—could catch and hold your pen the way streetcar tracks caught and held your bike's wheel; they threw you off your curve. But if you overcame this hazard, you could play at stretching and squeezing the Hogarthy face. I drew inside a textbook's illustrations, usually on the bare sky or on the side of a building or cheek. When I was very young, I sometimes drew on my fingernails, and hated myself for it.

I drew at home, too. My lines were hesitant. "You make everything out of hair," Amy complained. It was always faces I drew, faces and bodies, men and women, old and young, mostly women, and many babies. The babies grew as my sister Molly did; they learned to walk.

At Ellis, Molly was in the second grade. The little kids

didn't wear uniforms; she wore pretty dresses. I was a forward on the basketball team. Standing around in front of the school, I used to dribble Molly. She bounced hopping under my hand; we both thought it was mighty funny. During class, I drew her hopping in a smocked dress.

If I didn't draw I couldn't bear to listen in class; drawing siphoned off some restlessness. One English teacher, Miss McBride, let me sit in the back of the classroom and paint.

I paid no attention to the drawings. They were manneristic, obsessive, careless grotesques my hand gibbered out like drool. When I did notice them, they repelled me. Mostly these people were monstrous, elongated or compressed. Some were cross-hatched to invisibility, cross-hatched till the paper dissolved into wet lint on the desk. They were swollen of eyelid or lip, megalocephalic, haughty, moribund, manic, and mostly contemplative—lips shut, full-lidded eyes downcast, as serene as I was excited. They wore their ballpoint-pen hair every which way; they wore ill-fitting hats or melting eyeglasses. They wore diapers and ruffled pants, striped ties, brassieres, eye patches, pearls. Some were equipped with hands on which they rested their weary heads or which they waved, shockingly, up at me.

Very often I connected these unwittingly formed people by a pen line leading from the contour of a neck or foot to a drawing of the pen that drew the line and thence to my carefully drawn right hand holding the pen, and my arm and sleeve. I loved bending my thoughts down that pen line and up, that weird trail connecting and separating the conscious and unconscious: the wiggly face half-fashioned, and the sly, full-fashioned, and fashioning hand.

More than once, on family visits far away, or on the streets where I walked to school, or at Forbes Field, I saw a stranger whom I recognized. How well I knew that face, its bee-stung lips, its compressed forehead, its clumsy jaw! And I realized then, with a draining jolt of superstitious dread, that I was seeing in the flesh someone I had once drawn. Someone I had once drawn with a ballpoint pen inside a matchbook, or on an overcrowded page, a scribbled face inside the lines of a photographed woman's skirt. Now here was that face perfectly molded and fleshed in, as private as the drawing and as sad, walking around on a competent body, apparently experienced here, and at home.

Outside the study hall the next fall, the fall of our senior year, the Nabisco plant baked sweet white bread twice a week. If I sharpened a pencil at the back of the room I could smell the baking bread and the cedar shavings from the pencil. I could see the oaks turning brown on the edge of the hockey field, and see the scoured silver sky above shining a secret, true light into everything, into the black cars and red brick apartment buildings of Shadyside glimpsed beyond the trees. Pretty soon all twenty of us—our class—would be leaving. A core of my classmates had been together since kindergarten. I'd been there eight years. We twenty knew by bored heart the very weave of each other's socks. I thought, unfairly, of the Polyphemus moth crawling down the school's driveway. Now we'd go, too.

Back in my seat, I repeated the poem that began, "We grow to the sound of the wind playing his flutes in our hair." The poems I loved were in French, or translated from the Chinese, Portuguese, Arabic, Sanskrit, Greek. I murmured their heartbreaking syllables. I knew almost nothing of the diverse and energetic city I lived in. The poems whispered in my ear the password phrase, and I memorized it behind enemy lines: There is a world. There is another world.

I knew already that I would go to Hollins College in Virginia; our headmistress sent all her problems there, to her alma mater. "For the English department," she told me. William Golding was then writer in residence; before him was Enid Starkie, who wrote the biography of Rimbaud. But, "To smooth off her rough edges," she had told my parents. They repeated the phrase to me, vividly.

I had hopes for my rough edges. I wanted to use them as a can opener, to cut myself a hole in the world's surface, and exit through it. Would I be ground, instead, to a nub? Would they send me home, an ornament to my breed, in a jewelry bag?

I was in no position to comment. We had visited the school; it was beautiful. It was at the foot of Virginia's Great Valley, where the Scotch-Irish had settled in the eighteenth century, following the Alleghenies south.

Epilogue

A DREAM CONSISTS OF LITTLE MORE than its setting, as anyone knows who tells a dream or hears a dream told:

We were squeezing up the stone street of an Old World village.

We were climbing down the gangway of an oceangoing ship, carrying a baby.

We broke through the woods on the crest of a ridge and saw water; we grounded our blunt raft on a charred point of land.

We were lying on boughs of a tree in an alley.

We were dancing in a darkened ballroom, and the curtains were blowing.

The setting of our urgent lives is an intricate maze whose blind corridors we learn one by one—village street, ocean vessel, forested slope—without remembering how or where they connect in space.

You travel, settle, move on, stay put, go. You point your car down the riverside road to the blurred foot of the mountain. The mountain rolls back from the floodplain and hides its own height in its trees. You get out, stand on gravel, and cool your eyes watching the river move south. You lean on the car's hot hood and look up at the old mountain, up the slope of its green western flank. It is September; the goldenrod is out, and the asters. The tattered hardwood leaves darken before they die. The mountain occupies most of the sky. You can see where the route ahead through the woods will cross a fire scar, will vanish behind a slide of shale, and perhaps reemerge there on that piny ridge now visible across the hanging valley—that ridge apparently inaccessible, but with a faint track that fingers its greenish spine. You don't notice starting to walk; the sight of the trail has impelled you along it, as the sight of the earth moves the sun.

Before you the mountain's body curves away backward like a gymnast; the mountain's peak is somewhere south, rolled backward, too, and out of sight. Below you lies the pale and widening river; its far bank is forest now, and hills, and more blue hills behind them, hiding the yellow plain. Overhead and on the mountain's side, clouds collect and part. The clouds soak the ridges; the wayside plants tap water on your legs.

Now: if here while you are walking, or there when you've attained the far ridge and can see the yellow plain and the river shining through it—if you notice unbidden that you are afoot on this particular mountain on this particular day in the company of these particular changing fragments of clouds,—if you pause in your daze to connect your own skull-locked and interior mumble with the skin of your senses and sense, and notice you are living,—then will you not conjure up in imagination a map or a globe and locate this low mountain ridge on it, and find on one western slope the dot which represents you walking here astonished?

You may then wonder where they have gone, those other dim dots that were you: you in the flesh swimming in a swift river, swinging a bat on the first pitch, opening a footlocker with a screwdriver, inking and painting clowns on celluloid, stepping out of a revolving door into the swift crowd on a sidewalk, being kissed and kissing till your brain grew smooth, stepping out of the cold woods into a warm field full of crows, or lying awake in bed aware of your legs and suddenly aware of all of it, that the ceiling above you was under the sky—in what country, what town?

You may wonder, that is, as I sometimes wonder privately, but it doesn't matter. For it is not you or I that is important, neither what sort we might be nor how we came to be each where we are. What is important is anyone's coming awake and discovering a place, finding in full orbit a spinning globe one can lean over, catch, and jump on. What is important is the moment of opening a life and feeling it touch—with an electric hiss and cry—this speckled mineral sphere, our present world.

On your mountain slope now you must take on faith that those apparently discrete dots of you were contiguous: that little earnest dot, so easily amused; that alien, angry adolescent; and this woman with loosening skin on bony hands, hands now fifteen years older than your mother's hands when you pinched their knuckle skin into mountain ridges on an end table. You must take on faith that those severed places cohered, too—the dozens of desks, bedrooms, kitchens, yards, landscapes—if only through the motion and shed molecules of the traveler. You take it on faith that the multiform and variously lighted latitudes and longitudes were part of one world, that you didn't drop chopped from house to house, coast to coast, life to life, but in some once comprehensible way moved there, a city block at a time, a highway mile at a time, a degree of latitude and longitude at a time, carrying a fielder's mitt and the Penguin *Rimbaud* for old time's sake, and a sealed envelope, like a fetish, of untouchable stock certificates someone one hundred years ago gave your grandmother, and a comb. You take it on faith, for the connections are down now, the trail grown over, the highway moved; you can't remember despite all your vowing and memorization, and the way back is lost.

Your very cells have been replaced, and so have most of your feelings—except for two, two that connect back as far as you can remember. One is the chilling sensation of lowering one foot into a hot bath. The other, which can and does occur at any time, never fails to occur when you lower one foot into a hot bath, and when you feel the chill spread inside your shoulders, shoot down your arms and rise to your lips, and when you remember having felt this sensation from always, from when your mother lifted you down toward the bath and you curled up your legs: it is the dizzying overreal sensation of noticing that you are here. You feel life wipe your face like a big brush.

You may read this in your summer bed while the stars roll westward over your roof as they always do, while the constellation Crazy Swan nosedives over your steaming roof and into the tilled prairie once again. You may read this in your winter chair while Orion vaults over your snowy roof and over the hard continent to dive behind a California wave. "O'Ryan," Father called Orion, "that Irishman." Any two points in time, however distant, meet through the points in between; any two points in our atmosphere touch through the air. So we meet.

I write this at a wide desk in a pine shed as I always do these recent years, in this life I pray will last, while the summer sun closes the sky to Orion and to all the other winter stars over my roof. The young oaks growing just outside my windows wave in the light, so that concentrating, lost in the past, I see the pale leaves wag and think as my blood leaps: Is someone coming?

Is it Mother coming for me, to carry me home? Could it be my own young, my own glorious Mother, coming across the grass for me, the morning light on her skin, to get me and bring me back? Back to where I last knew all I needed, the way to her two strong arms?

And I wake a little more and reason, No, it is the oak leaves in the sun, pale as a face. I am here now, with this my own dear family, up here at this high latitude, out here at the farthest exploratory tip of this my present bewildering age. And still I break up through the skin of awareness a thousand times a day, as dolphins burst through seas, and dive again, and rise, and dive.

I GREW UP IN PITTSBURGH IN THE 1950s, in a house full of comedians, reading books. Possibly because Father had loaded his boat one day and gone down the Ohio River, I confused leaving with living, and vowed that when I got my freedom, I would be the one to do both.

Sometimes after dinner, when my sisters and I were young, Mother could persuade Father to perform Goofus—to "do" Goofus. Goofus, he explained, was an old road-show routine, older than vaudeville, that traveling actors brought to the cities of the young republic, and out into the frontier towns. It was a pantomime, a character, and a song.

Doing Goofus, Father shambled, holding his tall frame unstrung like one of those toy figures whose string collapses when you press the bottoms of their stands. He walked onstage a hayseed, a farm boy, a rube, sticking his neck forward. He sang the syncopated song, dipping his knees mock-idiotically on the beats:

> I was born on a farm down in Ioway,
> Flaming youth who was bound that he'd fly
> away . . .

Between verses of the song the rube stepped forward and concentrated on some absurdity, like balancing on his finger an imaginary hair plucked from his head. He stiffened the hair with his fingers and, wincing horribly, inserted it into one of his ears. He pushed it through to the other ear till he could grasp it; then he drew it sawing back and forth through his skull with both hands, in one ear and out the other, grinning stupidly in the character of the rube. That was the flaming youth.

"Do you know what you call that?" he asked as he sat back down. The intelligence had come back to his eyes. What? "That," he said. "What you do up there. That is called a 'business.' "

Our father taught us the culture into which we were born. American culture was Dixieland above all, Dixieland pure and simple, and next to Dixieland, jazz. It was the pioneers who went West singing "Bang away my Lulu." When someone died on the Oregon Trail, as someone was always doing, the family scratched a shallow grave right by the trail, because the wagon train couldn't wait. Everyone paced on behind the oxen across the empty desert and some families sang "Bang away my Lulu" that night, and some didn't.

Our culture was the stock-market crash—the biggest and best crash a country ever had. Father explained the mechanics of the crash to young Amy and me, around the dining-room table. He tried to explain why men on Wall Street had jumped from skyscrapers when the stock market crashed: "They lost everything!"—but of course I thought they lost everything only when they jumped. It was the breadlines of the Depression, and the Okies fleeing the Dust Bowl, and the proud men begging on city streets, and families on the move seeking work—dusty women, men in black hats pulled over their eyes, haunted, hungry children: what a mystifying spectacle, this almost universal misery, city families living in cars, farm families eating insects, because—why? Because all the businessmen realized at once, on the same morning, that paper money was only paper. What terrible fools. What did they think it was?

American culture was the World's Fair in Chicago, baseball, the Erie Canal, fancy nightclubs in Harlem, silent movies, summer-stock theater, the California forty-niners, the Alaska gold rush, Henry Ford and his bright idea of paying workers enough to buy cars, P. T. Barnum and his traveling circus, Buffalo Bill Cody and his Wild West Show.

It was the Chrysler Building in New York and the Golden Gate Bridge in San Francisco; the *Monitor* and the *Merrimack*, the Alamo, the Little Bighorn, Gettysburg, Shiloh, Bull Run, and "Strike the tent."

It was Pittsburgh's legendary Joe Magarac, the mighty Hungarian steelworker, who took off his shirt to reveal his body made of high-grade steel, and who squeezed out steel rail between his knuckles by the ton. It was the brawling rivermen on the Ohio River, the sandhogs who dug Hudson River tunnels, silver miners in Idaho, cowboys in Texas, and the innocent American Indian Jim Thorpe, who had to give all his Olympic gold medals back. It was the men of every race who built the railroads, and the boys of every race who went to war.

Above all, it was the man who wandered unencumbered by family ties: Johnny Appleseed in our own home woods, Daniel Boone in Kentucky, Jim Bridger crossing the Rockies. Father described for us the Yankee peddler, the free trapper, the roaming cowhand, the whalerman, roustabout, gandy dancer, tramp. His heroes, and my heroes, were Raymond Chandler's city detective Marlowe going, as a man must, down these mean streets; Huck Finn lighting out for the territories; and Jack Kerouac on the road.

Every time we danced, Father brought up Jack Kerouac, *On the Road*.

We did a lot of dancing at our house, fast dancing; everyone in the family was a dancing fool. I always came down from my room to dance. When the music was going, who could resist? I bounced down the stairs to the rhythm and began to whistle a bit, helpless as a marionette whose strings jerk her head and feet.

We danced by the record player in the dining room. For fast dancing, Mother only rarely joined in; perhaps Amy, Molly, and I had made her self-conscious. We waved our arms a lot. I bumped into people, because I liked to close my eyes.

"Turn that record player down!" Mother suggested from the living room. She was embroidering a pillow. Father opened the cabinet and turned the volume down a bit. I opened my eyes.

"Remember that line in *On the Road*?" He addressed me, because between us we had read *On the Road* approximately a million times. Like *Life on the Mississippi,* it was the sort of thing we read. I thought of his blue bookplate: "Books make the man." The bookplate's ship struggled in steep seas, and crowded on too much sail.

I nodded; I knew what he was going to say, because he said it every time we played music; it was always a pleasure. We both reined in our dancing a bit, so we could converse. Sure I remembered that line in *On the Road.*

"Kerouac's in a little bar in Mexico. He says that was the only time he ever got to hear music played loud enough—in that little bar in Mexico. It was in *On the Road.* The only time he ever got to hear the music loud enough. I always remember that."

He laughed, shaking his head; he turned the record player down another notch. If it had ever been at all, it had been a long time since Father had heard the music played loud enough. Maybe he was still imagining it, fondly, some little bar back away somewhere, so small he and the other regulars sat in the middle of the blaring band, or stood snapping their fingers, drinking bourbon, telling jokes between sets. He knew a lot of jokes. Did he think of himself as I thought of him, as the man who had cut out of town and headed, wearing tennis shoes and a blue cap, down the river toward New Orleans?

I was gaining momentum. It was only a matter of months. Downstairs in the basement, I played "Shake, Rattle, and Roll" on the piano. Why not take up the trumpet, why not marry this wonderful boy, write an epic, become a medical missionary to the Amazon as I always intended? What happened to painting, what happened to science? My boyfriend never seemed to sleep. "I can sleep when I'm dead," he said. Was this not grand?

I was approaching escape velocity. What would you do if you had fifteen minutes to live before the bomb went off?

Quick: What would you read? I drove up and down the boulevards, up and down the highways, around Frick Park fast, over the flung bridges and up into the springtime hills. My boyfriend and I played lightning chess, ten games an hour. We drove up the Allegheny River into New York and back, and up the Monongahela River into West Virginia and back. In my room I shuffled cards. I wrote poems about the sea. I wrote poems imitating the psalms. I held my pen on the red paper label of the modern jazz record on the turntable, played that side past midnight over and over, and let the pen draw a circle hours thick. In New Orleans—if you could get to New Orleans—would the music be loud enough?

ABOUT THE AUTHOR

ANNIE DILLARD is the author of ten books, including the Pulitzer Prize winner *Pilgrim at Tinker Creek*, as well as *An American Childhood*, *The Living*, and *Mornings Like This*. She is a member of the Academy of Arts and Letters and has received fellowship grants from the John Simon Guggenheim Foundation and the National Endowment for the Arts. Born in 1945 in Pittsburgh, Dillard attended Hollins College in Virginia. After living for five years in the Pacific Northwest, she returned to the East Coast, where she lives with her family.